GODS AND MEN

Gods and Men

Myths and Legends from the World's Religions

Retold by

JOHN BAILEY
KENNETH McLEISH, DAVID SPEARMAN

Illustrated by

DEREK COLLARD, CHARLES KEEPING,
JEROO ROY

OXFORD UNIVERSITY PRESS

OXFORD   NEW YORK   TORONTO

Oxford University Press, Walton Street, Oxford OX2 6DP

Oxford New York Toronto
Delhi Bombay Calcutta Madras Karachi
Petaling Jaya Singapore Hong Kong Tokyo
Nairobi Dar es Salaam Cape Town
Melbourne Auckland

and associated companies in
Berlin Ibadan

Oxford is a trade mark of Oxford University Press

A CIP catalogue record for this book is available
from the British Library

ISBN 0 19 274145 4

Printed in Great Britain

The stories of Odin, Beowulf, and the Quest for the Holy Grail are retold by David Spearman.

The Six Seeds, One Honest Man, Tseng and the Holy Man, Prometheus, St George and the Dragon, and Elijah and the Prophets of Baal are retold by Kenneth McLeish.

All the other stories are retold by John Bailey who edited the book.

CONTENTS

INTRODUCTION

Some of the earliest stories that we know are attempts to explain how the world was made, where mankind came from, and why so many things seem to be wrong with the world. These stories are called *myths*.

The word *myth* is sometimes used to mean something which isn't true. The way in which we are using the word here is quite different. When we call a particular story a myth, we don't necessarily mean that it is untrue. Of course, most of the myths you will find in this book are very old, and they were first told long before man began to think scientifically. So they will often contain details that we know today cannot possibly have been correct. But to the men who first told it, a myth was a way of expressing something which they firmly believed to be true, about their god or gods and their dealings with men.

Later, stories grew about great hero-figures of the past. Often these stories are about men grappling with the forces of evil in the form of monsters or giants, and succeeding either with the aid of their gods or by their own superhuman strength and cunning. These stories are called *legends*. Again, to call a story a 'legend' is not necessarily to say that it is untrue. Many legends are founded in historical fact; but they were handed on by word of mouth, and grew and developed in the telling.

The myths and legends in this collection come from different countries and religious traditions all over the world. These stories often bear striking similarities. The story of God becoming angry with the evil in the world and sending a flood to wipe out mankind occurs in the traditions of the Greeks, the Sumerians, the Jews, the Hindus, and the Chinese. In other ways, however, the stories differ sharply from each other. The Creation stories, for example, usually include the geographical features of their country of origin: the Scandinavian story of Odin is set among ice and rocks, whereas the Australian story of the Rainbow Snake tells of a hot, dry land where water was scarce.

As you read these powerful, ancient stories, think of the men and women who first told them to their children, and try to imagine their fears and doubts and the beliefs which they were expressing in story form in this way. How do these beliefs differ from the religious beliefs of men today? Ask yourself what *you* believe about the origins of the universe, the creation of man, and the problem of evil. And, like King Arthur's knights, keep searching.

John Bailey

CREATION MYTHS

Creation myths are stories told by men to answer the basic questions of life: how, did the world begin, and how did man come into the world? These stories go back to man's earliest history. They are not scientific, because they were told before man had begun to think scientifically.

Such stories are found in all cultures and all religions. They were handed down by word of mouth for hundreds of years before they came to be written down, and they were developed and often refined in the telling. They tell us much about the way of life of early man: his hopes and fears, his gods and demons, and the answers which he gave to the question of the meaning of life.

The Polynesian islands in the Pacific Ocean are the home of an ancient people with rich and varied religious traditions. The huge and mysterious stone heads of Easter Island are believed to be statues of gods, but little is known about them or the purpose they served. This story of Narreau the Elder is a creation myth from the Gilbert and Ellis Islands.

Narreau the Elder

Before anything was, there was Narreau the Elder. Nothing came before him; there was no animal, no fish, no bird, no man before Narreau the Elder. All around him was darkness and emptiness. There was no food, and Narreau neither ate nor felt hunger. There was no night and day, so Narreau did not sleep. For timeless ages, Narreau sat alone in the darkness.

Gradually, almost imperceptibly, Narreau began to change. You could not have picked out a single moment in time at which the change happened. But now, instead of being one, he was two: Narreau the Elder and Narreau the Younger.

The two gazed steadily at each other, and Narreau the Elder spoke.

'My work is nearly finished,' he said. 'All my knowledge and power is yours. There is one thing left for me to do; I will make a universe, upon which you shall practise your skills.'

So saying, Narreau the Elder stretched out a hand and created the universe. Then he slowly disappeared, like mists before the morning sun.

Narreau the Younger looked down at the universe that had been created. The earth and sea and sky were still fastened firmly together; they had not yet been separated. Thoughtfully, Narreau walked over the sky, looking for a way through; a crack, a crevice, any way to reach the land below. But there was none.

Narreau knelt down upon the top of the sky. 'I shall use my magical powers,' he said. 'Did not the Elder say that all his knowledge and power was mine? And did he not create this universe for me? I shall take what is mine.'

He stretched out his hand and tapped gently on the surface of the sky. Nothing happened. Narreau tapped again. At the third tap, the sky opened. Now Narreau could enter his universe.

He stood up and looked through the hole in the sky. All was pitch black, and he could see nothing. So he reached out both hands, and rubbed his fingertips together. There was a flash of light, and a little luminous moth appeared – the first creature.

Narreau smiled, and held out his hand. The moth fluttered down, and rested on his palm.

'With the light from your body, you can see through the darkness,' Narreau said. 'Go through this opening and tell me what you can see there.'

The moth flew down from the sky and disappeared from view. Some time later he returned, and again settled on the palm of Narreau's hand.

'There are people there, great Narreau,' said the moth. 'But because of the darkness, they are not moving – they are all asleep.'

'I shall see for myself,' announced Narreau. 'Lead the way, little moth, that I may see by your light.'

By the flickering light of the moth's luminous body, Narreau the Younger found his way to a low mound on the earth, in the middle of all the people. He stretched out his arms so that his fingertips brushed the sky.

'My people!' he cried out in a voice of thunder. 'I am Narreau, your Lord and Creator. I command you – move!'

At this, the sleeping bodies began to stir.

'Move, I say!' cried Narreau. 'Stand up, and lift the sky!'

The people began to stand, and the sky was lifted a little. Light began to appear, and the people blinked and rubbed their eyes.

'Higher!' called Narreau urgently. 'Help me to part the earth and the sky!'

The people pushed upwards, but the sky would not move

any further. Narreau realized that the sky was still firmly rooted to the land, and ordered them to rest. He looked around, and then called Octopus from his hiding place in the sea.

Octopus slithered his way across the land, and flopped down in front of Narreau.

'What is your command, oh Master?' he wheezed.

'Return to the sea, Octopus, and fetch Conger Eel from the depths. Tell him I have a task for him that will make him great for ever.'

Conger Eel, the mighty Lord of the deep sea, could not resist this appeal to his vanity. There was a great surge of foam, and he appeared. His great body curled, he waited for Narreau to speak.

'Greetings, Lord of the Depths,' said Narreau. 'Reach up with your mighty body, lift up the sky with your head, and press down the land with your tail. The time has come for the sky and the land to be separated.'

Eel uncoiled his huge body and pressed up against the sky. Slowly, the great roof of the sky moved upwards as its roots were torn out of the land. The land sank and more light appeared. Narreau looked up; the sky was far enough above the land, but there were no sides to it.

'I shall pull down the sides of the sky, and complete my world,' Narreau said. Leaping up, he caught hold of the edge of the sky and pulled it down to the land, while Eel kept the centre supported on his powerful body. Narreau leapt again and again, pulling down the edges of the sky until it was securely fastened all round and the sky was shaped like a bowl. A shadow fell across the land, and Narreau the Elder reappeared.

Narreau the Younger leapt again, and with one thrust of his magical sword his father lay dead. Narreau the Younger took his father's right eye and flung it into the sky to the East. It curved through the sky, lighting up the world with dazzling brightness; it was the Sun. Narreau took his father's left eye, and flung it West. It was the Moon. Narreau took the fragments of his father's shattered body and threw them into the sky, where they became the stars.

Turning to the land, Narreau planted a tree from which

grew men; the ancestors of people today. They were the Kings of the tree of Samoa, the red-skinned people with the blue eyes.

At last Narreau's work was complete. He looked at his creation, the work of his hands, and as he looked the Sun began his journey across the sky. The light grew, and the outline of Narreau the Younger's body slowly became misty and vanished, as his father's had done before. Narreau was never seen again.

*Although the Australian continent was not discovered by explorers
from the Western world until the seventeenth century, men had been
living in Australia for some ten thousand years. This story of the
spirit god and his son, Chinimin, is one of the myths of these original
inhabitants, or Aborigines. This story reminds us of the importance
of rain in a country where water is scarce.*

The Rainbow Snake

In the spirit world, the spirit god reigned supreme. All the
spirits were afraid of him, for was he not Lord of All?
And then, one day, the unthinkable happened. Someone
dared to disagree with the spirit god. It was Chinimin, the
god's son, who had spoken out, and the god was furious.
Chinimin must be punished!

In a rage, the spirit god rushed at Chinimin, who turned in
fright and ran to the river which separated the spirit world
from the earth. He hoped that if he crossed it he would escape
his father's anger. But the spirit god saw what Chinimin was
trying to do, and sent a swarm of bees to head him off in
mid-stream.

Half suffocated by the bees and tossed about by the water,
Chinimin was driven back to the river bank where his father
was waiting. At once, the spirit god, huge, gnarled, set about
him with a club. Soon the spirit world echoed to the clash of
the battle between the old god and the young Chinimin.

Chinimin seized a spear, and evading his father's blind
fury, thrust it home. The spirit god, badly wounded,
doubled up, convulsed with pain, and fell to the ground.
Writhing and twisting in agony, he pulled himself along,
desperately trying to seize hold of something to stop the pain,
something to give him strength to fight back. He dragged
himself to the bank of the spirit river, his body contorted into
the shape of a gigantic snake.

The waters of the spirit river parted as the snake-god in his

torment crashed from the bank. Hissing with pain, the god fell from his spirit home to our planet, Earth, which at that time was featureless and dry. He pulled with him water from the spirit river. A huge crater formed where his snake-body fell to earth, and was filled with the water from the spirit river, forming a vast lake.

The force of the spirit god's fall was so great that it created a huge whirlwind which roared and spiralled across the flat, dry surface of the earth. Sobbing and racked with pain, the snake god followed the whirlwind in search of an end to his desperate suffering.

As the great snake's body dragged along, it formed hollows and valleys in the landscape, and raised up hills and mountains. The water from the spirit river rolled along the valleys behind him, forming streams and rivers. From time to time he paused, his body coiled up in pain. When he forced himself to move again, deep hollows were left in the ground; they filled with water and became lakes.

The water from the spirit river gave life to the land, and plants sprang up in a profusion of colours – red from the blood of the great snake, blue from the water of the spirit river, white from the clouds in the sky, yellow with the glint of the sun which now shone down upon the earth through the hole in the sky left by the spirit god's fall. Gradually, as he struggled on, the body of the great snake took on these colours.

The meeting of clouds and sun, cold and hot, began to make storms. The dark clouds rolled together; thunder growled and lightning cracked. Rain lashed down, till the new rivers flooded and burst their banks.

As the spirit forces of thunder and lightning, sunshine and rain, racked the earth, mysterious stirrings began. Life was being formed. The energy of the storm and the sunshine, the water from the spirit river, and the blood of the rainbow snake combined to cause animals, insects and birds to appear and flourish. Finally, as the last clap of thunder died away and the rain stopped at last, Man himself was born.

Now at last the wounds of the spirit god in the shape of the rainbow snake began to heal. He looked up at the sky, to his real home beyond the spirit river. Coiling his body, he sprang up through the clouds and returned to the spirit world, leaving his rainbow-snake skin sloughed and hanging in the sky above the world.

He left behind an earth completely transformed. What had been arid desert was now a watery paradise, filled with colour and new, thrusting life.

African religion is rich in myths and stories of gods, and many stories of Nyame the sky-god are to be found among the peoples of West Africa.

Fire Children

Nyame the sky-god lived above the darkness of the African night sky, alone in space. One day he took a basket, filled it with earth and plants, placed all kinds of animals and insects inside it and hung it up in the sky. That basket is the earth. He made a trapdoor in the sky, so that he could climb down to visit the earth, and holes so that it would not be completely dark on earth when the trapdoor was shut. The trapdoor and holes in the sky are the moon and stars.

One day, when he looked down from his trapdoor, Nyame noticed that part of the earth was bare. He liked to see plants and flowers blooming everywhere, so he filled another basket with growing things and lowered it to earth on a rainbow-rope.

All this time there were spirit people living inside Nyame, and just as he was leaning out of his trapdoor to lower the basket to earth, two of them decided to climb up inside him and look out through his mouth to see what he was doing. Just as they reached his mouth, Nyame sneezed and blew them out. They fell down through the trapdoor, down through the sky, and landed on the earth.

What a strange place it was, after the home they were used to! The spirits wandered among the trees and plants, fascinated by the colourful blossoms and the funny little animals they saw. They found a warm, sheltered cave to live in, and settled down to a new life on earth. They missed their safe home inside Nyame, but they could think of no way of getting back up to the trapdoor.

Often, the man spirit would leave the woman spirit inside the cave, and go and talk to the wind and trees, dance with shadows and the falling leaves. The woman spirit, whose name was Iyaloda or Great Mother, used to get very lonely. But she was afraid to go with him, and she could not persuade him to stay all day with her.

At last she thought of a way to end her loneliness. One day, when the man spirit returned to the cave, he found her full of excitement.

'What is it, Iyaloda of many moods?' he asked. 'When I left, you were grumpy and sad. Now you glow with excitement like a hibiscus-flower.'

'I've made a plan,' Iyaloda answered, 'a plan to end my loneliness forever.'

'You've made a plan!' said the man spirit angrily. 'Don't you remember the last plan you made, when we were living happily inside Nyame, the great one? It was your idea to climb up to his mouth to see what he was doing – and if it wasn't for that we'd still be safe inside him, instead of being stranded here on earth!'

'Listen,' said Iyaloda. 'Just listen to the plan. It can't hurt just to listen.'

'All right,' said the man spirit sulkily. 'What is your plan?'

At once, Iyaloda began to explain. 'If we take some clay and mould some small creatures, shaped like ourselves,' she said, 'we can bake them in the fire and then breathe life into them so that they can move about like us. Then they'll be here to keep me company, and I won't be lonely when you go off to the forest. We could call them children.'

She waited to see what her man would say. He was full of relief that this plan was so simple, so straightforward; and so he agreed that they should try.

They built up the fire till it was glowing red, and carefully modelled the first batch of little figures with clay which the woman spirit dug up outside the cave. They put the figures in the fire to bake, and sat back to wait.

Just then they heard the footsteps of the great sky-god Nyame crashing through the forest, and heard his voice

calling their names. Terrified in case he found them doing something they shouldn't, they snatched the little clay children out of the fire and wrapped them in leaves to hide them.

They were just in time. Nyame appeared at the mouth of the cave and stood looking down at them.

'Well, spirit people,' he boomed, 'how do you like my earth? Have you been good?'

'Oh yes, lord Nyame,' they replied, bowing before him.

'Well, see that it continues,' the sky-god said. 'Take good care of all I have made, and above all, keep out of mischief.'

He left them, and walked off to enjoy his world. As soon as he was gone, they made a new batch of clay-people, and put them in the fire to bake. But they had hardly got them in before Nyame appeared again. This time they had to leave the little figures baking and hope that the sky-god wouldn't notice. He seemed suspicious this time, because he stayed a long while; but he went away without discovering the secret.

The sky-god called to see them several more times during his visit to earth. Sometimes they heard him coming, and took a batch of children out of the fire quickly; at other times they had to leave the little figures baking until he was gone.

At last Nyame went back to the sky, and the spirit people were able to take out all the children they had made and look at them. Some were hardly baked at all, and were quite white. Others had turned yellow; some were baked red or brown; a few were burned quite black.

The spirit people were delighted with all the children they had made, whatever the colour. They breathed their breath into each of them, so that they came to life, just like children waking up from sleep.

Iyaloda, the Great Mother, was lonely no longer.

Off the East Coast of Africa lies one of the largest islands in the world, Madagascar. Its people are descended from both Africans and Indonesians, and this story of their god Driana-nahary tells us how and why he came to create men and put them on the island.

Sand Children

High up in the clouds, the god Driana-nahary looked down on the hot, sun-baked, empty land that we now call Madagascar, and wondered what could be done with it. Could any living being survive down there? Eventually he made up his mind. There was only one way to find out – someone must go down and see.

Driana-nahary looked round for his son Atoko-loinona. Where was the lad? Then he saw Atoko on another cloud, playing happily with a ball of lightning. Driana-nahary smiled. He was fond of his son, but being a great and important god, he used to pretend to be stern with him.

'Atoko! Come over here!' he called.

Atoko-loinona put down the lightning and ran across the springy clouds to where Driana-nahary stood.

'You called me, Sir?' he said. Atoko was a polite and obedient boy.

'I've got a job for you, lad,' said Driana-nahary.

'Yes, father?' said Atoko eagerly. To be quite honest, he sometimes got a bit bored up in the clouds, although he wouldn't have dared tell his father so.

'Now, this is important,' Driana said. 'There must be no playing about on the way. You see that land down there?'

He pointed through a gap in the clouds. Atoko peered down and saw the land spread out, with the shadows of the clouds moving slowly across it.

'I want you to go down and have a look,' said his father.

'See what it's like down there. It looks hot and dry, and I need to know whether anything could survive down there. Off you go – and come straight back.'

Atoko was delighted. He had often wondered what it would be like down there on earth. He went over to the gap in the clouds, waved goodbye to Driana-nahary, and let himself slide down over the edge.

An ordinary human boy would have been frightened to fall to earth from such a great height. But Atoko was a god, and he floated gently down and landed lightly on his feet. At once he could feel that the earth, so beautiful from above, was not so wonderful after all. The ground was hot from the sun; it scorched the soft soles of his feet, and Atoko looked round desperately for some shade or some water to cool his feet.

All around him was a blank, sandy desert. No plants, no trees, no shelter from the sun, and, worst of all – no water. Surely nothing could live down here. Atoko made up his mind at once to fly back to his father and make his report. He climbed to the top of a nearby sand-dune, and leaped up, ready to fly. But instead of soaring upwards, he fell in a heap in the hot sand.

Spluttering and coughing the sand from his face, he sat up. What had happened? He couldn't fly. Again and again he jumped, and again and again he sprawled on his hands and knees in the sand. What was he to do?

Sobbing, limbs heavy as lead, Atoko-loinona dragged himself across the desert, seeking vainly for some way of escape. The sun beat down; he could feel his strength slipping away. Just before he reached the point of collapse, Atoko summoned up all his reserves of power, and burrowed deep inside the ground to hide from the scorching heat.

Meanwhile, in the clouds, Driana-nahary was getting worried. Night fell, and still there was no sign of his son. Driana stayed by the gap in the clouds all night, waiting for Atoko, and when morning came he was nearly frantic with anxiety.

'What have I done?' he cried. 'Atoko! Atoko! Where are you? Come back!'

But the earth lay empty and silent. Atoko-loinona was nowhere to be seen.

There was only one thing for Driana-nahary to do. Whatever the risks they faced on earth, he must create living beings and send them down to look for his son. Swiftly, he made an army of men and placed them on the earth's surface with instructions to search everywhere for the missing boy.

The men fanned out like ants in all directions over the hot desert; but they could find no trace of Atoko. Before long, they too were in great distress. They were mortal, and they needed food and drink to survive. But there was no water, no plants, no trees. They sent messengers to shout up to Driana-nahary in the clouds and ask for fresh instructions; but the messengers failed to return.

Up in the sky, Driana-nahary sat heedless, head in hands. How could he have done such a thing – sent down his only son? Now the boy was missing – perhaps gone forever. Then another thought came to him. His army of servants – how would *they* survive on the hot, dry land? He looked down, and there they all were, spread out, still searching for Atoko-loinona. Some lay still, bent like withered leaves. The others moved slowly, as hunger and thirst dragged at their strength.

Driana realized he must do something or they too would perish. He reached out and struck the clouds a great blow with his hand. There was a flash of lightning, a clap of thunder, and the clouds burst, sending cooling, refreshing rain flooding down onto the land.

Far below, the men Driana had created looked up in gratitude as the rain began to fall. They cupped their hands to catch the water, and drank. At first, the raindrops hissed and spat as they hit the scorching ground. Then, as the water soaked into the surface of the earth, plants and bushes began to spring up, and the desert was transformed into a paradise.

Years passed. Driana-nahary's ant-army, mankind, settled down on earth, learnt to grow plants and crops for food, married and had children.

Atoko-loinona was lost forever under the earth, never seen again. But Driana-nahary, although grief lay always in his heart like a stone, had a new race of children to care for, his new creation, the race of men.

Egypt was the home of one of the world's earliest civilizations. The first written records of Ancient Egypt go back to about 3200 BC, and this civilization lasted for over three thousand years. The Egyptians worshipped many gods, one of which was the Sun, which they believed to be a form of the creator god, Ra. This is their story of creation.

Ra and Sekhmet

Before Egypt, before the world, before time itself, there was only Nun, god of water, and his son Ra. Ra was the god of creation: he had only to think of a thing, and at once that thing was born.

'I will be sunrise, the blazing sun at noon, and last, the sunset,' he said. No sooner had the words passed his lips than he was transformed into the sun, rising in the east and sinking in the west. The first day was created: time was born.

Next Ra thought of the wind, and named it Shu. He thought of rain, and named it Tefnut, the spitter. He thought of dry land rising above the sea, and named it Geb, or Egypt. At once land appeared from the waters, the wind blew gently over it, and rain began to fall. Ra created a beautiful goddess, whose body arched across the earth. She was the sky; her name was Nut. Then Ra made the river Nile, to flow through the land and bless it with plants and crops.

One by one, Ra thought of all the inhabitants of earth: plants, animals, birds, insects, fish. As each came into his mind, it was created and named. His last creation was mankind: he made a man and a woman, and before long there were people living all over Egypt.

Ra's people needed a leader. So he himself took the form of a man, and ruled Egypt. His name was Pharaoh. He ruled for thousands of years, and Egypt was peaceful and prosperous. The people worshipped Ra; god of the sun, eye of the day; it was the world's golden age. Every year the Nile flooded so that crops would grow on either side. There was plenty and happiness; the people were content.

Because Ra had taken the form of a mortal man, he began at last to grow old as all men must. When this happened, many of his people turned from him, and began to worship Apophis, the dark one, the spirit of evil.

Ra sent for the gods he had made: Shu god of the wind, Tefnut the rain, Geb the earth and Nut, goddess of the sky. He even sent for Nun, the god of water, the oldest of the gods, from whom he himself was born.

The gods gathered before their king, the sun. He gazed at them with his fiery eyes, and said, 'Nun, my father, and all you other gods, my creation: look at my people, the men and women I made and placed in Egypt to be my servants. They have turned from me; they worship Apophis, the spirit of evil. His poison fills their hearts. What shall I do? With one glance of my burning eye, shall I destroy them all?'

'No!' said Nun. 'Why destroy them all? Why kill good and bad alike? Why make Egypt, that lovely land, a scorched desert? Create a power, an avenging spirit that will kill the bad, and leave the good alive.'

'What you say is just,' said Ra. 'I will create Sekhmet, the destroyer!'

As soon as the thought was formed and uttered, Sekhmet appeared. She was a lioness, the size of a house, with sharp teeth and rending claws. Her one longing was to feed on the blood of men. She bounded across the countryside, seeking out evil, hunting down Apophis' followers.

Thousands were caught and killed. The Nile foamed red, and the land along its banks was choked with carcases drained of blood.

Soon, all the wicked were slain. But now Sekhmet had the taste for human blood. She turned on the good people too. Terror walked the land; the people cried to Ra for help.

Ra called his fastest runners, and sent them to the island of Elephantine to fetch red ochre, a powerful dye found nowhere else. While they were gone, he set every slave in his palace to work making barley beer.

By the time the runners returned, every room in the palace was crammed with clay jars of beer. 'Mix the ochre with the

beer,' ordered Ra. 'Mix it carefully, till it looks like blood.'

That night in the darkness, a long column of Ra's servants left the palace, each with a jar of blood-red beer. They made their way to the fields near Sekhmet's lair. In the silent night they could hear her slobbering and snoring as she slept, glutted with slaughter.

The men poured the beer over the fields, smashed the empty jars, and hurried away.

Next morning, Sekhmet awoke thirsty as usual. She padded out from her greasy lair, foul-smelling with death, and prowled through the countryside looking for men to kill.

She came to the blood-red fields. Her thirst dulled her sense of smell, and she lapped and gulped the oozing red beer till the fields were brown and dry again.

But this blood was different. Her head felt muzzy and thick, and her legs staggered under her. She flopped in a heap on the river-bank, and gazed ahead with unfocussed, boozy eyes.

There was a sudden radiance, a dazzle of light. The Sun himself, Ra the creator, had come upon her.

'Peace, Sekhmet,' he said. 'From now on you are Sekhmet the Slayer no more. I give you a new nature, and a new name: Hathor, goddess of love. You will still have power over men, even greater power – but now it is the power of love, not hate.'

So the evil of Apophis was defeated by Ra the shining one. The wickedness in man was punished, and the good preserved. The power of hate was changed forever: love was born.

Many of the myths of the North American Indians tell of a series of worlds that existed before this world. This account of creation comes from the Navajo Indians.

Coyote the Trickster

In the First World there were only three beings: First Man, First Woman, and a creator-god known as Coyote the Trickster because of his cleverness. These three moved to the Second World, where the Sun and the Moon lived. But there, Sun fell in love with First Woman and this led to many arguments between him and First Man. In the end they all decided to climb to the Third World where there would be room for Sun to separate from First Woman.

Third World was a spacious and beautiful land, rather like Earth today. It had a mountain at each of its four corners, and a huge lake at the foot of each mountain. Living on the slopes of the mountains the newcomers found a race of friendly people, who made them welcome and told them all about their new surroundings. Third World, they said, was a happy and peaceful place – just so long as you didn't disturb Tieholtsodi the water-monster.

First Man and First Woman were impressed by the tales that the mountain people told about the water-monster, and resolved to keep well away from the lakes. But Coyote was curious, and went off to the eastern waters to investigate. Lying on the banks of the lake, he came across two beautiful children fast asleep under a tree. They were the monster's children; but Coyote picked them up, wrapped them in a blanket, and hurried off home with them.

When the monster discovered that two of his children were missing, he was furious. He splashed up and down the shallows of the eastern lake, searching everywhere for them.

Then he set off to the other three corners of the world, roaring with fury. But his children were nowhere to be found. The monster guessed that the newcomers must have something to do with it, and he resolved to use his power over water to punish them.

Soon the waters of the four lakes began to rise, and the land was flooded. The mountain people could see that if the water continued to rise, even their mountains would soon be flooded. They resolved to pile the four mountains up, one on top of the other, in the centre of the world, and escape the flood that way.

Laboriously, the mountain people shifted the huge mountains to the middle of the world, and climbed up to the top. They perched there, watching wide-eyed as the waters continued to rise. Soon the rippling flood was only a few feet below them and they could hear Tieholtsodi the water-monster chuckling and bubbling in the depths as he took his revenge.

As a last resort, the mountain people planted a giant reed on the mountain-top and waited anxiously as it grew up towards the sky. It touched the sky and pierced through into the Fourth World. First Man and First Woman, Coyote the Trickster with his blanket-bundle containing the monster's children, the mountain people and all the animals from the Third World scrambled inside the reed. Last of all came Turkey; he was to gobble out an alarm when the flood-water reached him.

The flood continued to rise; Turkey's tail got wet, and he gave the warning cry. They all began to climb up inside the giant reed, towards the sky. Even now turkeys have light-coloured tail feathers to show where their ancestor was washed by the flood.

When they climbed out of the top of the reed, they found that the Fourth World was a place of mist and emptiness, with great blocks and mountains, seas like the Third World and a central plain divided by a swift river.

Everyone settled in the Fourth World, and for a while all was peaceful. Then men and women began to quarrel. Each claimed that their work was the most important. The men argued that they did the hunting, built the houses, and that they knew the ways of the gods and performed the dances and ceremonies that made the crops grow. The women said that they tended the crops, cooked the meals and looked after the children.

The quarrel grew so bitter that in the end men and women decided to separate. The men built a boat and poled it across the river, leaving the women behind.

Four years passed. The men became more and more resentful that they had to labour in the fields tending their crops with no hot meals to go home to at the end of the day. On the other side of the river the women found that their crops were dwindling, because there was no one to work the land. At last, realizing that men and women need each other after all, they joined forces again.

But Coyote the Trickster still had the children of Tieholtsodi, and the water-monster had not given up. The waters rose, the ground of the Fourth World became soft and marshy, and again the people were threatened by floods. Once more they tried to escape by planting a giant reed and climbing up inside. This time, Badger went first; but at the top he found nothing but a muddy lake. No-one knew what to do, and at the bottom of the reed the water was still rising. Eventually Locust, with a great effort, found a way to hold his breath and swim up through the marshy lake. Gasping and spluttering, they all followed him. Each carried a bundle of their most precious possessions; Coyote's bundle was a pair of bulging blankets.

Just as they reached dry land and began to congratulate one another on their escape, what should appear in the marshes but the horns of Tieholtsodi himself. He had caused the flood in the Fourth World to reach up and fill the marshes of the Fifth World. Now he roared and splashed till the sky bellowed fear. White with terror, everyone opened their bundles so that he could search for his missing children.

Coyote the Trickster was caught at last. He had to hand over the monster children, and the people watched in relief as the water-monster and his babies swam away. Gradually the waters followed him, trickling down to the underworld. First Man, First Woman, Coyote the Trickster, and all the others were left at the Place of Emergence in the Fifth World.

As the flood waters drained away, they saw that they were standing on a small island separated from the mainland by a swamp. They prayed to the gods, and gradually the swamp dried out sufficiently for them to cross in safety. The mainland was rough, unfinished mud, oozing and slithering in shapelessness. To anchor it, the people piled stones and mud into mountains at the four corners. Then they threw Sun and Moon up into the sky.

At first Sun was too near the earth. For four days the earth expanded like a soap-bubble in the heat; Sun flew up and out, higher and higher. At last it stood still at its highest point, above the centre of the earth. It grew hotter and hotter, until everything began to wither and scorch. The people huddled together wondering what to do.

As they clustered together in the meagre shade, the wife of one of the great chiefs gave a sudden cry and fell to the ground. Everyone gathered round, and her husband cradled her head in his lap. But her life's breath ebbed away and she died. As her body grew cold, it slowly vanished – and at once Sun began to move again across the sky. Coyote explained to the people that every day someone, somewhere in the land must 'die', so that the Sun could keep moving.

This caused great amazement, because no-one knew what 'die' meant. No one had ever 'died' before. Was it a game? A trick? They didn't know whether to be alarmed, or pleased. All they knew for certain was that the chief's wife had disappeared.

While all this was going on, one man, who had made his way back to the Place of Emergence, looked down through the hole in the bottom of the world and saw the dead woman sitting happily by the river of the Fourth World combing her hair. She told him that after death, everyone must return from the Fifth to the Fourth World to live.

The wise man returned to the others and told them what he had seen and heard. Shortly afterwards, he too fell sick and died. Coyote told the people that death would come quickly to anyone who looked upon the face of the dead. From that time on, the faces of dead people were covered until they were buried; and no Navajo ever followed the wise man's tracks and ventured to look for the Place of Emergence.

A story of the creation of monkeys and men, from the mountains of Guatemala in Central America.

Hurakan and Gucumatz

At the beginning of time, all things were under the water, except the givers of life, Hurakan and Gucumatz. They hovered over the water like giant humming birds. When the time came, Hurakan and Gucumatz said the word 'Earth', and slowly land was formed and rose up out of the water.

The creators saw that the land was flat, and caused hills and mountains to appear. Then they flew over the landscape that had been created, and the earth was at once covered with vegetation.

Next, Hurakan and Gucumatz made animals and set them down on the earth. They commanded the animals to worship them, but the animals could not understand or speak. They grunted and snuffled, roared and howled, but Hurakan and Gucumatz were not satisfied. They decided to make men more intelligent than the animals. These new people would know their creators and offer the praise and worship they wanted. Hurakan and Gucumatz debated what to do with the animals, and eventually decided to leave them on the earth so that men could kill them and eat them for food.

First, they modelled some human figures out of clay and placed them on the earth. But the clay men could not move or speak or understand; they were worse than the animals. Hurakan and Gucumatz could think of no way of making the clay men come to life, so they destroyed them.

Next, they carved some wooden men and stood them on

the land. The wooden men looked perfect, and were much stronger than the clay men; being made out of living wood, they could move about, but they still had no intelligence or feelings and could not recognize their creators. Hurakan and Gucumatz decided to destroy them and start again, but this time some escaped and became the monkeys that we see today in the forests.

Hurakan and Gucumatz talked for a long time about the problem of making men who would be intelligent enough to worship their creators, but not so intelligent that they would want to overthrow the gods. Eventually they decided to make four men out of the life-giving plant, maize. They took strands of the yellow and white maize and carefully twisted and plaited them into the shapes of four men. This time, they were successful. The men they had made were perfect; they could run, jump, think, talk, eat and drink.

Indeed, in one way, they were too perfect; their eyesight was so good that they could see Hurakan and Gucumatz quite clearly, high up in the heavens. The gods held no mystery for them when they could be seen so easily, and Hurakan and Gucumatz realized that if they wanted the men to worship them, they would have to do something about their eyesight. So they waited until the men were asleep, then carefully shortened their sight so that they could no longer see into the heavens. Now the gods appeared dim and mysterious to the men, and they began to worship their creators as Hurakan and Gucumatz desired.

The creator gods were so pleased with the worship of the four men that they made four women for them, to be their helpers and companions. So human life on earth had begun.

An ancient Middle Eastern civilization, that of Sumeria, flourished in Mesopotamia from earliest times until about 2000 years B.C. The story of Marduk is the Sumerian account of creation.

Marduk

Anu, ruler of the heavens, had a son called Ea, the god of knowledge. Ea knew everything that there was to know. Every being in the world contained Ea; he knew their secrets and all their thoughts.

Ea lived in a palace high on Mount Lebanon, with his wife, the goddess Damkina. It was there that their son Marduk was born. Ea was so proud of his son that he gave him twice as much power as all the other gods. Marduk had two pairs of eyes, two mouths and four ears. Like many spoiled children, he was always up to mischief. He used his eyes and ears to spy on the other gods; he was forever spreading gossip and playing jokes on them.

Many of the gods took no notice. 'He'll soon grow out of it,' they said. But others took his teasing badly, and decided to kill him. They went to Tiamat, goddess of the sea, for help.

Tiamat listened to their complaints. Then she cackled with glee, and said, 'I'll do what I can. But remember: his father has given Marduk twice as much power as any of us. Our own strength won't be enough. I must make an army of monsters. Then we can win. My monsters will soon bring Marduk to heel.'

At once Tiamat began. She made fire-breathing dragons, scaly monsters with teeth and horns, spiny giants with tails like scorpions. Some glowed red like a sullen fire; others were milky, indistinct as ghosts. When the army was ready, Tiamat set out to find Marduk and deal with him.

The gods were in council when news of the monsters came. One by one they peered through the shutters at the advancing horde; one by one they crept back to their places, trembling with fear. Not even Ea the all-knowing knew what to do.

At last one of the gods stood up and said, 'Enough! What are we afraid of? Tiamat and her monsters have no quarrel with us. It was Marduk who offended them. Let Marduk go out and deal with them.'

There was a babble of argument. Some said he was right, that it was Marduk's fault, and Marduk's bad luck. Others said he was just a boy, too young to fight. The discussion went back and forth for hours; outside, the army of monsters came nearer and nearer.

At last Marduk was sent for. When they told him, his answer took them all by surprise. 'Of course I'll fight,' he said. 'But there's one condition. Since I'm the only one of you capable of defeating Tiamat, you must make me your king. Do that, and I'll fight for you.'

The gods grumbled and groaned. A boy, their king? But it was fair enough. Boy or not, if Marduk defeated Tiamat, he deserved to rule.

Marduk, king of the gods, was armed for the fight. A royal crown was placed on his head, and the king's sword by his side. The royal war-chariot was brought in, its horses snorting and pawing the ground. Marduk fetched weapons of his own: a bow and a single arrow; storm-clouds, and lightning that made his chariot flash with fire. And finally, a net, a fishing-net strong enough to trap a monster in.

When he was ready, he climbed into the chariot and rode out across the plain. The storm-clouds billowed round his head; the lightning crackled at the horses' heels; the golden crown glittered in the sun.

The monsters saw him, and stopped in their tracks. They were huge and fierce; but they were also new-born, and knew nothing of war. They whimpered and cowered. But Tiamat gathered them and gave them strength. She breathed defiance into them, until they stood round her like a hideous hedge, snorting and terrible. Behind them, well sheltered, the rebel gods watched, eager for Marduk's death.

Marduk knew that Tiamat was his only real enemy. She was the others' strength; she gave them their power. If he killed her, their fight would wither away and die.

But how could he reach her? Surrounded by demons, she was safe.

Then a plan, a last mischievous trick, came into his mind. Instead of challenging Tiamat, he began mocking her, laughing at her and shouting names. For a while Tiamat stood her ground; but at last, quivering with fury, she shouldered her monsters aside and rushed at Marduk to kill him hand-to-hand herself.

As she hurtled past, Marduk side-stepped. He flung the net over her head, and tangled her. The more she struggled, the more she was trapped. He took his bow and his one arrow, and shot her through the heart. Tiamat was dead.

When they saw Tiamat fall, the monsters and the rebel gods panicked and began rushing away in all directions. But Marduk knew what to do. He unharnessed his storm-winds and sent them to round them up; then he chained them and hurled them into the underworld.

The forces of evil were defeated; now Marduk was truly king of the gods.

When the loyal gods saw Marduk's victory, they streamed out of the palace to meet him, singing his praises and acclaiming him king. At once he began by creating a new kingdom for himself, and setting it in order. First, he took a club and broke open Tiamat's skull. Her blood was carried by the storm-winds over all the universe, like sap, taking new life wherever it flowed. Next, he split her body in two. Out of one half he made the sky, out of the other the earth.

From among the loyal gods, Marduk appointed Anu to rule space, Enil to be lord of the earth and sky, and Ea to command rivers and the sea.

Lastly, he studded the heavens with stars, ordered the sun to orbit the earth, and hung the moon in the night sky. From then on, time was constant: days, seasons, years, all settled. Order had come upon the world.

To prevent the gods grumbling at the work they had to do, looking after the newly-made world, Marduk decided to create a servant-race. 'They will have no power,' he said.

'They will be weak, no threat to us gods. Their place will be to do all the menial tasks that are not for us. To prevent them learning our skills in time, or growing strong enough to challenge us, I will make them mortal: after a certain time, they will be bound to die. Their name will be Man.'

To create man, Marduk used the body of one of the defeated gods. He cut off the immortal head, and out of the trunk a new creature was fashioned: mis-shapen, ridiculous, a poor copy of the immortal gods. Stumbling and shuffling, the new creature crept off to begin his lowly life. Man was born.

The gods were delighted with Marduk, their king. In his honour, they built a glittering temple in Babylon, city of gods and men. Here Marduk ruled in splendour. Every new year's day gods and men gathered in Babylon to worship Marduk their king, the creator who ordered the universe and brought it peace.

The ancient Norsemen, ancestors of the people who today live in the Scandinavian countries of Denmark, Sweden, Norway and Iceland, told many stories of a race of super-men descended from gods and giants. The greatest of these was Odin, and this story tells how Odin and the gods came to create the world.

Odin

ICE AND FLAME

Before time began, there was nothing. Emptiness . . . a black void, a shuddering chasm swirling with mist and spray, torn by the groans of heaving, cracking ice. On one side, banks of dank mist, cold as corpse-breath. On the other fire, scorching, ravenous.

Cold and heat, ice and flame, invaded each other . . . came together . . . mingled.

From the hissing and blistering of their mingling two Beings were born: Ymir, first of the giants, and Audumla the gentle cow.

The baby giant found the cow's udder, and sucked strength. He grew and grew . . . limbs like oaks, breath like a wind soaring in the pines. And as he grew, new giants sprang up from the drops of his sweat and the soles of his feet.

Audumla the cow licked the ice-blocks in the void, eager for the salty taste that was in them. Caressed by her rough tongue, the aged ice softened . . . thawed . . . trickled away. After one day her licking loosened hairs of a head, frozen in the ice-block. Another day, and the head itself was free. Three days, and the god Buri burst from his prison, stretching his cramped limbs till the last slivers of ice splintered and fell away.

In the black, swirling ice-chasm Buri, first of the gods, confronted his enemy, the giant Ymir. Each wanted to rule, to be first. And the first step to power, the beginning of time, would come when one killed the other.

On one side Buri: huge, alone, powerful. On the other Ymir and his frost-giants, snapping and cracking white fingers of cold. All round, the writhing mists of space, billowing and whispering round the silent, panting battle-ground.

For a hundred centuries, a thousand ages, the battle ebbed and flowed. The noise of it roared and rumbled through space like a hundred thunderstorms together. Neither side could win: giants and god, they were equally matched.

Then at last the god Bor, Buri's son, married a giantess. Their children, god-giants, giant-gods, were Odin and his brothers. They had the power and strength of both giants and gods: they were invincible.

Odin and his brothers fought on the gods' side. At last they struck down the huge giant Ymir. Blood poured from the ruined caves of his body, a red tidal wave that engulfed and choked the other giants. They floundered in the red death, and drowned. Only two escaped; flopping and splashing in the sticky red sea, they swam with their last strength to the outermost rim of space, and clung gasping on the edge like stranded fish. When their strength returned they built a dark tower, a fortress of stone and iron. Swaggering safe behind its walls, they shook knotted fists at the darkness of space, and vowed vengeance on all the race of gods.

Led by Odin, the victorious gods took Ymir's body and hurled it into the empty void. From it they formed our world. His body became earth, his blood the seven seas; from the dome of his skull they fashioned the curving sky, and his brains made drifting clouds. From his eyebrows they wove a bristling hedge, to protect earth from the simmering anger of the outcast giants.

Bones for mountains, teeth for rocks, hair for vegetation . . . The work was almost done. Only the maggots crawling in his dead earth-flesh were left. These they changed into elves: bent, evil elves to work with metals under ground; shining elves to tend the fair gardens of the world above.

Finally, Odin reached his hand into the glowing, searing furnace on one side of the void of space. He took handfuls of

living sparks and flung them upwards, to glow like fireflies in the night sky. Two of the largest became the sun and moon. They would have stayed anchored in their places forever, but in their dark iron fortress, the jealous giants became aware of them, and sent a pack of grey wolves panting across the roof of the world, to hunt the sun and moon and keep them in terrified motion.

For themselves, the gods created a kingdom, Asgard. It was high above the earth, and joined to it by a many-coloured rainbow, a royal road. In Asgard they met in council, drank their mead and took their ease. When they chose, they went strolling in the lower world, and their presence was a thunderclap, an eruption, a sudden shudder of majesty and power.

Below, on middle earth, flowers opened for the first time and the sea whispered and boomed on the empty beaches. But there was no one save elves and gods to savour its splendour and freshness, no one to take delight in the newness of creation.

One day Odin and his two brothers were walking idly by the shore. They found two logs of driftwood, and on an impulse whittled them into godlike shapes: wooden dolls. Each god breathed a gift into the rigid forms: warm life, beauty – and lastly, flowing movement. Before them, tiny on the seashore, stood Ask and Embla, the first man and woman in all the world.

THE REBIRTH

Mighty Odin knows that for all his power and his victories over the booming giants and their allies, he and his fellow gods will be destroyed one day. This is their fate, which even they cannot avoid.

First, three terrible winters will take hold on the world. Each will follow the other without a break, so that the dying earth has no time to recover between. Ice will grow thick and hard as steel, locking the earth in its steely arms, squeezing it like a workshop vice. The sun will weaken; every day its rays will grow more feeble, until at last the terrible tireless wolves, slavering and snarling across the sky, will catch the sun and moon, tear them into shining pieces and gulp them down.

When that moment comes, the brooding giants in the dark tower at the edge of the universe will bellow in triumph. The stars will sputter out, like dying sparks flung from a burning log. The black void will return. Mountains and rocks will shudder, and loosen their roots like rotten teeth. In the sea, the monster Iormingard, who has grown continually since time began, will know it is time for him to rise and move his swelling coils towards the land. So violent will be the motion of his colossal length, like the steel spring of a gigantic clock, that the sea will be flung in torrents over the land, and drown it.

Worse will follow. There is a cunning god, Loki, who has played many tricks on the others. Through jealousy he caused the death of the sun-god Baldur, and all this time he has been clamped between flat rocks in a cavern below the earth as a punishment. Above him, a venomous serpent drips poison on to his face. Now at the end of time, his chains will snap like tinsel, and he will join his monster children in the last battle.

Loki's children are Fenris Wolf, stronger than anything in creation, till now fastened to a rock by magic cord made by the elves; Iormingard the sea-monster; and Hel the giant-maiden, queen of the underworld. All the giants of forest and storm, quivering and burning for vengeance, will join with them against the gods.

On a tower on the battlements stands Heimdall, watch-man of the gods. His eyes are keen as a hunting eagle's and he sees in the distance the hordes of evil appear, gathering like a swarm of flies against the gods. Heimdall takes his horn, and blows a blast that rings round the halls of Asgard and echoes through the bright air, to warn the immortals with golden sound.

The gods hurry to defend themselves. The spirits of the dead warriors in Valhalla stream out to join them. Swords clink and rattle, as cups of mead crash over in their haste, and glittering winged-helmets are thrust on their shaggy heads. Odin himself, king of the gods, has a helmet of crusted gold that flashes like a mirror to dazzle his enemies. This he puts on and rides out at the head of his army.

On the other side, the forces of evil gather. Like a thick polluted fog, they pour over the rainbow bridge that arches across space and links the corners of creation. It crashes in ruins under the burden of giants and demons, monsters and scowling elves – but not before they spill, howling and shrieking, into the gods' own kingdom of Asgard.

Now the last battle begins. Thor, swinging his great hammer, attacks the serpent and kills it. But even as he does so, he himself falls dead from the creature's dank, sulphurous breath. Odin, the mighty Odin, hacks and thrusts at Fenris Wolf. But the wolf is too strong even for him, and swallows him whole: the beast devours the god. Soon the wolf is destroyed in its turn, its jaws wrenched apart by Vidor. Loki and Heimdall strike each other dead in a single blow.

So the battle fares. At last, scorching fire bubbles from the void and roars into ruined Asgard. Everything is destroyed – gods and champions, monsters and giants, bodies, buildings, even the tree of life itself.

Yet through this destruction, the world will be cleansed. The prophecy is that the last grinding battle will not be the end of existence. The Sun has bred a daughter, even brighter than herself. After a long period of darkness a new earth will arise from the sea. The sun god Baldur will return from the dead; the young surviving gods will make their home where Asgard was, and find again the golden chessmen with which their fathers played. A new race of men and women will people the earth. Evil will have disappeared, and happiness will prevail in all the universe.

Hinduism, the ancient religion of India, has many gods and many myths and legends. Most of the gods of Hinduism are seen as forms of two central Gods, Vishnu and Siva. This story tells of the creation of the world by the God Brahma at the command of Vishnu.

Manu and Shatarupa

Before the world, before the sky, before space, there was nothing but ocean: a flat, rolling lake that lapped the edges of emptiness and the void beyond. Floating on the water was a giant water-snake: Ananta, Serpent-King. In his coils, eyes closed, undisturbed, lay the Lord Vishnu. God, asleep. Water, snake, god: nothing moved. Stillness . . . perfection . . .

Then in the deepest recesses of the world, a sound began. A slow gathering, a humming, a throbbing. It grew and pulsed and filled the emptiness: a power, an urge, a throbbing itch of energy. It billowed and gathered into a single echoing syllable, folding in on itself endlessly, endlessly, like a beating heart: OM . . . OM . . . OM . . .

Lord Vishnu opened his eyes. It was time. The world was ready to be born.

He looked out over the calm waters. In that moment, a lotus flower took shape before him. In it sat Brahma the Creator, the Lord Vishnu's servant. He bowed his head, and waited to hear Lord Vishnu's will.

'It· is time, Brahma. Time for the world. Time to begin your work. In that single lotus flower is all you need. Create a world that will live forever, till I declare the end of time itself. Begin.'

As he spoke, a huge wind gathered. The ocean cowered. The Serpent-King, and Lord Vishnu with him, disappeared from sight. Alone, Brahma's lotus-boat tossed in the churning sea.

Brahma raised his arms, and the wind died. The sea fell back and was calm again. He stood up, and with a sweep of his arms divided the lotus into three parts. The first part was heaven, the next earth and the next sky. In a single moment, the world had begun.

Brahma clothed the new earth with plants: grass, trees, flowers, vegetables and fruit. To them he gave the sense of touch. Then he created animals and insects – large and small, in land, sea and air, some with fur, some with feathers, some with shells, some with scales; large and small, fierce and timid, fast and slow. To them, as well as the sense of touch, he gave sight, smell, hearing – and above all, the power of movement.

At once the world filled with flurry and bustle. With crashing of branches, clatter of hooves, swishing and swooping, flailing and flapping, the new creatures set off to find homes. Trumpeting, braying, whistling, chattering, squealing, they ran and wriggled and hopped and flew into every corner of creation.

In the stillness that was left, Brahma had only one thing more to do. The world needed a master, someone to enjoy it and take care of it, so that it would last forever, as Lord Vishnu had commanded. Brahma sat quiet, and thought. After a long time his thoughts took shape. First, a wisp of shadow in the white air . . . a glowing, shimmering cloud that grew thicker and denser, solidifying into a living, breathing shape. A new being, made from the thought of Brahma, in the form of god. Brahma looked at him in delight: surely this creature, made in god's image, would take charge of the world and keep it forever as Lord Vishnu wished.

But the creature did not move. Its eyes were shut, unheeding the new world around it. Because it was made of the thoughts of Brahma, all it wanted was to sit thinking deeply about god.

Brahma saw that his creature was too simple, too flawless to look after the world. If he was to create a being to carry out Lord Vishnu's will, he would need another power. Thought was not enough: he would need to use action too. Not only his mind, but his whole body, his whole self, would be required if the new creature was to open his eyes to the world, be happy and fulfilled by creation as well as the creator.

There was only one certain way. Filled with contentment that he was carrying out Lord Vishnu's orders, Brahma divided his own body in two. One moment there was one, the next there were two: equal, unblemished, whole, the image of one another. Out of one, Brahma shaped man; out of the other, woman. The man was called *Manu*, wise; the woman *Shatarupa*, mysterious.

Manu and Shatarupa, created out of Brahma himself, looked into each other's hearts. They smiled. Gently, they touched hands. Then they walked out together into the world Brahma had given them; their charge, their responsibility, the joy and the duty laid on them by Lord Vishnu at the start of time.

Manu, Shatarupa . . . the first people . . . the ancestors of the whole human race.

GOOD AND EVIL

If an all-powerful God made the world, why is it not perfect? Where did evil come from? Stories which attempt to explain the existence of evil, suffering and death, often connecting them with man's wilful disobedience, are found in many different religious traditions.

One particular theme occurs time and again: that of the Creator God, angry with man for his failure to obey, deciding to wipe out the whole of mankind with a flood.

Why do the seasons exist? Are the seasons in nature like those in men's lives, the seasons of birth and death? Modern science gives one answer. The Greek legend of Persephone gives another.

The Six Seeds

In the beginning there were no seasons. There was light and dark, sun and rain, heat-wave and frost. But the year needed no order, no pattern, In one field men would be ploughing and sowing seed; in the next would be standing corn, ready for harvesting. Demeter, goddess of the earth, looked after each growing thing and nursed it to ripeness in its own due time.

Demeter's daughter was Persephone. Of all the spirits and nymphs who lived on earth, she was the most beautiful. The meadow-flowers, poppies, anemones and asphodel, opened their petals to welcome her; tall trees, willows and poplars and cypresses, bent their heads for love of her; even rocks and lakes seemed to smile when she passed by.

Persephone's favourite place in all the world was the island of Sicily. Here, in sunny meadows by placid streams, she walked with her friends the nymphs. In the cool of the day they ran and played, brushing the meadow-grass with their finger-tips; when it was hot they sat in the shade, laughing and singing till evening came. They were like a vision, a dream of happiness for all the world.

Down below, in the caverns of the underworld, there was no light, no happiness, no living thing. In the cold, chill darkness fluttered wraiths of the dead, gaunt spectres numberless as fallen leaves. They had no memory, no future hope. They waited, not knowing what for; they ached with a pain they could not feel.

Only their king, Hades brother of Zeus, knew what they

missed. He remembered the light, the blossom of the upper world. When the three brothers divided creation, light and life and supreme power went to Zeus; Poseidon was given the sea; to Hades went the kingdom of the dead. A mighty lord, with riches and power equal to Zeus himself, he was sour with envy, bitter with longing for the light.

When the longing grew too much, he harnessed black horses to a dark chariot and galloped out across the world above. As he passed, mortals cowered. His black cloak billowed and hid the sun; his horses' eyes glowed red like coals; his dark chariot hissed with a hiss of death.

Hades rode on, like a storm-cloud passing across the earth. He came to Sicily, and saw Persephone playing with the meadow-nymphs. Her beauty filled his heart, like sun in an empty room. He was warmed by her; he ached for her. Here at last was a queen to share his kingdom, a goddess to bring light and hope to the world of the dead.

In the meadow, the sunlight was suddenly blotted out. The nymphs looked up and screamed. A dark shadow hovered over them, wings rustling, like a giant bat; there was a moment's chill, a whisper of cold, a shiver of death. Then it was gone. The sun shone again, birds sang, the stream murmured beside the reeds.

They looked round for Persephone, and screamed and screamed again. Where she had been, the grass was flattened and dead. Slivers of ice glittered among the fallen poppy-heads. Persephone had been snatched away. Only her belt, embroidered with flowers, was left, torn and faded on the frozen ground.

At first, in the underworld, Persephone cried and cried. What use was Hades' love to her? He was icy cold and terrible: his touch made her shake with fear. What use were his riches? Diamonds, rubies, emeralds: rock-treasures, cold and dead. What use was being queen? Her people were ghosts, husks drained of life.

Then, slowly, she began to change. Instead of ice and cold, she saw Hades' loneliness. She felt that her people, wraiths though they were, loved and needed her. She began to see beauty even in the lifeless jewels, the caverns of the under-world. Her heart ached for the flowers, the light, the warmth

of the upper world. But now she saw that her beauty, there, had been an ornament; here, it was a necessity, to give her people life and hope.

For the people of the upper world, all life and hope were gone. When Persephone disappeared, her mother Demeter was filled with cold despair. Day after day she wandered the world, twisting Persephone's belt till it was frayed and threadbare, endlessly calling her daughter's name. The plants and growing things were forgotten. All over the fertile world crops withered, shoots died, seed lay rotting in the ground. The starving people wept, and begged Demeter to pity them. But she could think of nothing but Persephone.

Finally, as famine gnawed away their strength, the people lifted their hands and prayed to Zeus for help. He heard their voices, thin as waves on a distant shore. He knew where Persephone was. He had seen his brother Hades snatch her away; he had seen her unhappiness, and watched it fade as she grew to accept her place as queen of hell.

But now the dead were balanced against the living. The wraith-people of the underworld needed Persephone; but the mortals of the upper world depended on her too. The scales were even: the choice was for Zeus to make.

He sent for Demeter, and issued his command.

'Your daughter rules in the underworld. And in the underworld she has eaten food: six pomegranate seeds. It is the gods' law that if a living soul eats food of the dead, it must stay forever with the dead. But I have seen the suffering of mortals in the upper world, and balanced it against the suffering of the dead. This is my command. For each seed she has eaten, Persephone must stay one month each year in the underworld, ruling with her husband Hades. For the rest of the year she is free to live in the world above.'

And so it was. For six months each year Persephone rules in hell. The upper world lies cold and bare, waiting for her return. At last she comes, and brings life to growing things. The months go by and soon, after harvest, it is time for her to return again to the kingdom of shadows. For us, summer is a time of happiness, winter a time of grief. For the wraiths below, our seasons are reversed: they suffer in summer, and rejoice in winter when their queen comes back to them.

This Jewish myth of the creation of man by the Lord Yahweh, and man's 'fall' from his original state of perfection, is central to three major religions of the world: Judaism, Christianity and Islam. This version is based on chapter two of the first book of the Old Testament, Genesis.

The Garden of Eden

When Lord Yahweh first made the world, it had no plant life growing anywhere because there was no rain, and there were no men to do any planting. However, there was a flood every so often, when the water surged out from underground and covered the whole land, so that the earth was ready for planting.

One day Lord Yahweh formed a man from the mud left after a flood, and breathed life into him. Yahweh called him Adam – the Hebrew word for 'man'. Realizing that Adam would have to have somewhere to live, Yahweh planted a garden in the East and called it Eden. He made all kinds of trees and plants spring up, so that the garden was sheltered to live in and beautiful to look at, and there was plenty of fruit for Adam to eat. Right in the middle of the garden, Yahweh planted two special trees – the Tree of Life and the Tree of Knowledge. Anyone who ate the fruit from the Tree of Life would live for ever, and anyone eating the fruit of the Tree of Knowledge would immediately be able to tell the difference between right and wrong.

When the garden was ready, Yahweh set Adam down gently in the garden to look after it.

'Here you are – this is your home,' Yahweh told him. 'There are lots of good things to eat, but keep away from that tree in the middle of the garden. That is the Tree of Knowledge, and if you eat any of its fruit, you will die.' At that time, although Adam was a fully grown man, he was as innocent as a baby, and went about naked.

After a while, Yahweh realized that Adam was lonely in the great garden all by himself. First of all, Yahweh shaped all the wild animals and birds out of the mud, brought them to life, and let Adam choose names for them all. But still Adam had no-one to talk to. Yahweh decided he would have to make a mate for Adam.

That night, while Adam was asleep, Yahweh took a rib from his chest, healing the place over as he did so. Then he used the rib to make a woman. When Adam woke up, he saw the woman and exclaimed:

'At last! A mate for me. Because her body was made from mine, and I am *man*, she shall be called *woman*.'

One of the cleverest of the animals in the garden was the snake. One day, he was talking to the woman.

'I hear you're not allowed to eat any of the fruit from the trees in the garden,' he hissed slyly.

'Oh yes, we can,' the woman replied innocently. 'We can eat any of the fruit we like, except the fruit of that tree in the middle of the garden. The Lord Yahweh told us that if we eat that, we shall die.'

'Of course you wouldn't die,' the snake scoffed. 'That's the Tree of Knowledge, silly. If you were to eat the fruit of that tree, you'd become a god, like Lord Yahweh himself. That's why he doesn't want you to eat it.'

The woman looked at the fruit on the Tree of Knowledge, and it really did look quite harmless. The more she thought about it, the more tempted she was. She went up to the tree, put out her hand – then pulled it back in fear.

'Go on – what harm can it do?' hissed the snake.

The woman reached out again, took an apple from the tree, and hesitantly took a bite. It seemed harmless enough, just as the snake had said. She took some to Adam, and he tasted it too. At first, nothing was different. Then, gradually, the fruit began to take effect. The first thing they realized was that they were both naked; it hadn't mattered before, but now, suddenly, it seemed wrong, so they made themselves simple loincloths out of leaves.

Just then, as the evening breeze began to rustle through the trees, they heard Lord Yahweh walking through the garden towards them. The man and woman looked at one another

guiltily. What had they done? Surely Yahweh would find out
and punish them. Adam took the woman's hand and they ran
and hid among the trees.

'Adam! Where are you?' called Yahweh.

Sheepishly, Adam came out from the trees and walked
towards Yahweh.

'Adam – what's the matter?' asked Yahweh. 'Why were
you hiding?'

'I heard you coming, and I was frightened because I hadn't
any clothes on, so I went and hid behind a tree,' said Adam
lamely.

'Clothes?' said Yahweh sternly. 'Who told you about
clothes? Have you been eating the fruit of the Tree of
Knowledge – the fruit I expressly told you not to eat?'

'It was the woman's fault!' said Adam accusingly. 'That
woman you made to be my mate! She gave it to me.'

Yahweh looked at the woman, who by this time had come
out from her hiding place and was standing next to Adam
with downcast eyes.

'What have you to say for yourself?' Yahweh asked her.

'It was the snake – he tricked me, and I ate the fruit,' the
woman answered.

Yahweh drew himself up and looked scornfully down at
the unhappy pair. They shuffled their feet and waited for
Yahweh's verdict. Yahweh saw through their flimsy
excuses. First of all, he addressed the snake.

'You miserable creature! Because of this act of yours, you
shall be different from all other animals from now on. You
shall crawl and slither about on your stomach, and eat dirt for
the rest of your life. Everyone will hate and mistrust you –
particularly this woman and her children. They will strike
out at your head whenever they see you, and you will strike
at their heels.'

Then Yahweh turned to the woman.

'Your punishment is even worse. When you have children,
it will be agonizingly painful for you. And you will not be
able to avoid having children.'

Finally, Yahweh spoke to Adam.

'There is no excuse for your disobedience. You listened to
the woman, and ate the forbidden fruit. Because of you, I

curse the land on which you are standing. No more will you live a life of leisure here in the garden. If you want to eat, you will have to dig and sow, and struggle to grow food for the rest of your days. Thistles and thorns will choke your crops, and you will have to work on the land until the day you die and are buried. You were made from dust, and to dust you will return.'

Because the woman was to become the mother of the human race, Adam called her Eve – Hebrew for 'life'. Yahweh made clothes for Adam and Eve out of animal skins, while he pondered what to do with them.

'Man has become like one of the gods – he has the power of knowledge,' he mused. 'What if he were to eat the fruit of the Tree of Life as well? Then he would live for ever! I cannot let them stay here in the garden – it is too dangerous.'

So Lord Yahweh drove Adam and Eve out of the Garden of Eden for ever, and barred the entrance to the garden with sharp swords so that mankind could never return.

At some time between 1500 and 600 B.C. – there is much argument among scholars about the correct date – a prophet known in the Iranian language as Zarathustra lived and taught in Iran. The Greek form of his name is Zoroaster, and Zoroastrianism became the religion of Iran for over 1,000 years.

Zoroaster believed he had been called by the god Ahura Mazda, Lord Wisdom, the Spirit of Goodness. He taught that man must choose between the goodness of Ahura Mazda and the evil of Ahriman, the Spirit of Evil. This is the Zoroastrian myth of creation.

Ahura Mazda and Ahriman

Ahura Mazda, the Spirit of Goodness, lived in the Heavens in endless light. He knew all things and was aware that across the void, in the very deep of dark, dwelt a Spirit of Evil, Ahriman. For many years, the two lived separate existences without coming into conflict. Ahriman, in his darkness of ignorance, did not even know that Ahura Mazda existed.

When Ahriman first saw Ahura Mazda surrounded by dazzling brightness, his evil nature compelled him to attack. Ahura Mazda offered him peace if he would worship goodness, but Ahriman thought that such an offer must be made from a position of weakness; he refused, and swore to destroy all that was good. Ahura Mazda saw that if the battle went on for ever, evil would eventually win, so he suggested to Ahriman that the battle should last only twelve thousand years. Ahriman was stupid as well as evil; he agreed to the suggestion and so missed the chance of victory.

After Ahura Mazda had fixed the period of battle, he recited a prayer so powerful that Ahriman fell back into Hell and remained there for three thousand years. Thereupon Ahura Mazda began to create. Out of the light, he created heavenly beings, the 'Immortals' and the 'Venerable Ones'. Ahriman meanwhile, in the depths of hell, created an army of demons to oppose the Immortals.

Ahura Mazda created the sky, which became a shell to enclose the world and also a prison for Ahriman. One of the

Immortals, Ameretat, caused ten thousand different plants to grow – and to one of them, the Ox-Horn Tree, he gave the magical power of giving eternal life to all who ate of it. Wicked Ahriman at once tried to destroy the Ox-Horn Tree by sending a lizard to attack it, but Ameretat protected it by putting ten good fish in the sea which surrounded it and causing them to swim round and round so that one of them was always watching the lizard.

The creation of animals came next. First, Ahura Mazda made a gigantic ox, but it was a senseless, clumsy beast that raged over the earth doing no good at all. Then the god Mithra was born out of a solid rock. Mithra appeared fully grown, with a torch in one hand and a knife in the other. He clothed himself in fig leaves, then made a treaty of friendship with the sun. Looking around, Mithra saw the ox, and seizing its horns, sprang upon its back and tamed it.

At this point a raven appeared with a message from Ahura Mazda to Mithra. He was to take his knife and kill the ox. The young god obeyed at once, and from the dead body of the ox sprang up all species of animals.

Just as the death of the ox gave rise to the birth of all other animals, so the death of one man brought about the birth of all mankind. For three thousand years the spirit of the first man, Gaya Maretan, had existed in the world with the mighty ox, and at last Ahura Mazda gave him a body formed out of his own sweat. For thirty years, Gaya Maretan struggled against a thousand demons sent to destroy him by Ahriman. Finally Ahriman succeeded in poisoning Gaya Maretan, and from his body came various metals, including gold. One of the immortal spirits carefully preserved the gold for forty years, whereupon it turned into the first human pair, Mashya and Mashyoi, the parents of the human race.

The descendants of Mashya and Mashyoi were intended by Ahura Mazda to live in his kingdom and be on his side in the long struggle between goodness and evil, but as time went by many joined Ahriman instead and became evil. But the battle will end; a saviour will be born, whose name will be Soshyant, and good will finally triumph over evil.

*There are many stories in the Hindu Scriptures concerning the
activities of the gods. Here is a story of Vishnu taking human form as
a little boy, Krishna, and coming into conflict with the forces of evil
embodied in the serpent Kaliya. It is taken from the great Hindu
epic, the Mahabharata.*

Krishna and the Serpent

Vishnu the preserver, lord of the universe, took the
form of a little boy called Krishna, so that he could
fight against the evil in the world. Krishna was dark-
skinned and handsome, and was the ring-leader in all the
mischievous games the village children played. His mother
and the other women laughed at his pranks; Krishna was
everyone's favourite.

When he was seven, Krishna and his half-brother Rama
used to go with the cowherds and help look after the cattle.
All day long they played in the fields and woods where the
cattle grazed, making garlands of leaves and flowers and
running in and out of the trees. Krishna was musical, and he
often sat down and played enchanting tunes on a flute he had
made. The other children and the cowherds sat at his feet, and
even the cows, and the birds and animals of the forest, came
closer to listen to the magical sounds.

The cowherds usually took their cattle to the banks of the
river Kalindi to quench their thirst. But one day all the cows
that drank from the river fell ill and died. The evil serpent-
king Kaliya had entered the river, and his presence poisoned
the water.

Soon nothing could live in or near the river. The fish died;
birds flying over the water were scorched and burned; even
the crocodiles left the riverbanks and crashed through the
forest in search of fresh water.

No one knew what to do. Everyone was afraid. At last

Krishna decided that the time had come for him to find the serpent in its lair and kill it. He set off alone, and went upriver till he came to a deep pool where the water foamed and boiled, and the trees were dead along the banks – Kaliya's lair.

'It was to overcome the wicked that I came into the world in human form,' Krishna thought. 'For the sake of my people now, I must dive into the serpent's home and kill him.'

He climbed a dead tree on the riverbank, edged out along a branch, and dived into the water.

Kaliya was furious at being disturbed by a little human boy, and rushed at Krishna to kill him. Krishna wriggled out of reach and swam to the surface to breathe. Kaliya followed him, trying to seize him in his coils and drag him down into the depths. Soon the water was seething with the serpent's poison and the writhing of its body. Krishna seized its head and climbed on to it, carefully keeping clear of the flickering, poisonous fangs.

Angrily, Kaliya tried to shake him off. Krishna hung on for his life. The water roared and churned round him; the snake wriggled and plunged; there was a ringing in his ears as the poison entered him from the water, and slowly took effect. The world went dark.

By now, some of the cowherds had missed Krishna, and followed him up river. They arrived just in time to see Kaliya dragging the boy's unconscious body down into the swirling depths. Horrified, they ran to fetch Rama and the other villagers.

Soon hundreds of frightened people lined the banks. They peered down into the murky depths. In the slime and mud of the river-bottom, Krishna lay still and silent in Kaliya's coils. The people cried out and wrung their hands in grief; many of them loved him so dearly they were ready to jump into the poisonous waters to try and save his life.

But Rama held them back. He was Krishna's half-brother, and knew of his divine nature. He looked down into the dark water, and called out in a loud voice: 'Krishna, great lord of all the gods, have you forgotten who you are? You are Vishnu the preserver, lord of the universe. It was to overcome evil that you became Krishna. Don't give way to

human weakness; use a god's power, and crush the serpent!'

At the bottom of the river, Krishna heard; he opened his eyes and smiled. Using a god's power, he flexed his body to break the serpent's grip, then sprang free. He jumped on to the monster's head, and began to dance. All his music, all his godlike skill, poured out in a dance of death on Kaliya's head. Gradually, Kaliya was forced to submit. He stopped struggling, and at last lay dazed and still.

Now Kaliya's wives, the serpent queens, saw that their king was beaten. They came out of hiding, and begged Krishna to spare his life. 'We did not recognize you, lord of the gods,' they cried. 'Now we see you are Vishnu, lord of all. Be merciful. His poison is spent. Spare his life, and let him go.'

The little boy Krishna swam to the bank, and looked down at the beaten monster.

'I created you, Kaliya, and now I will spare your life,' he said. 'Go now, with all your wives. Never again enter the waters of the river Kalindi. Your home is the ocean; stay there, and harm no mortal man again.'

So, with his wives and followers, Kaliya the serpent-king left the river Kalindi and swam to the ocean. The river waters became clear and pure again. The villagers cheered and praised the little boy who was also their lord and god.

Often people think they are the only honest people in the world, and that everyone else, thanks to dishonesty, is better off. The religions of the world tell many stories (including this one from China) to prove the opposite: that goodness and honesty, in the end, will always pay.

One Honest Man

Once there was a poor man named Yo. He was as honest as sun in summer, and that was why he was poor. All round him, people prospered: lies and trickery were the road to wealth, and it seemed as if the gods favoured only wicked men.

One day Yo was resting at an inn. He sat alone, ignored by the others. They roared and laughed, swaggered and snapped their fingers. They took no notice of Yo, quiet in his corner.

The door opened and a giant of a man, tall as the room, came in. He threw himself down next to Yo and stared gloomily ahead of him. All round, the others laughed and drank; Yo and the stranger were like a pebble of stillness on a beach of noise.

After a while, Yo asked the stranger if he was hungry. He pushed across his own plate, with his own supper on it. At once the giant picked up the food in handfuls, and wolfed it down till there was nothing left.

Astonished, Yo called for the landlord and ordered another dinner. That, too, was wolfed down like the last. Yo called for the landlord again. He gave him every coin he possessed, and ordered a shoulder of pork and two dozen boiled dumplings. Like the other food, these were soon disposed of.

When his meal (enough for half a dozen men) was done, the stranger sat back. He looked at Yo, who was politely waiting. 'Thank you,' he said, in a voice like boulders rumbling in a swollen stream. 'I haven't eaten a meal like that in three long years.'

'But why is that?' asked Yo. 'A fine big man like yourself – why should you lack for a job, or food to fill your stomach?'

The giant replied, 'The will of the gods is not to be discussed.'

Yo saw that he must change the subject. 'Where are you from?' he asked.

'On land I have no house, in water no boat. At dawn I live in the village, at night in the city,' the stranger replied.

Later, when Yo got up to leave, the stranger got up too. 'Let me go with you,' he said.

Yo and the stranger walked along the quiet roads in the still night. They talked of many things, and Yo was surprised at the stranger's knowledge and wisdom. He was especially knowledgeable about everything to do with water: clouds, rain, rivers and the sea itself.

The stranger stayed with Yo for many weeks. He ate only every twelve days, but each time he put away enough food for half a dozen men.

Now Yo, as has been said, was a poor man. He enjoyed the stranger's company and did not begrudge him the food he ate. But the stocks in the store-room were falling fast and it would not be long before new supplies were needed.

Yo had no money to buy food in town; everything depended on a good harvest in his two small fields. Every day Yo went out to look at his fields and his neighbours' fields, and see that all was well.

But Yo's neighbours were as wicked and unjust as all the others. They lived in a dry region, where water was scarce. By skilful trickery, changing the water-courses in the middle of the night, they diverted what little irrigation there was from Yo's fields to their own. Their crops grew green and tall; his were withered and brown.

As the dry days wore on, Yo looked out across his parched fields and sighed. 'If only I understood how rain begins,' he said to the stranger. 'If only I could walk among the clouds and see for myself.'

The stranger said nothing. It was the heat of noon. Yo went inside, and lay down to take a nap.

After a while, Yo felt a floating, spinning sensation. He opened his eyes and found he was lying in a fleecy white

substance, curling and twisting like smoke. He jumped up in
alarm, and the whiteness rocked and stirred under him. His
feet sank into it and he stumbled giddily about. He stretched
out his arms for balance and his right hand touched a hard,
smooth object, like a pebble suspended in the air.

It was a star. The purple-blue canopy above his head was
studded with stars, set like seeds in a lily-cup. Some were the
size of pudding-basins; others were tiny, like glowing coals
suspended in space.

Yo picked one of the tiniest and hid it in his sleeve. Then
he bent down and parted the clouds underneath his feet. He
looked through the gap and saw, far below, the glittering sea
and the land spread like a patchwork quilt. The houses were
no bigger than beans; the people looked like specks of dust.

Just at that moment, a pair of dragons with scaly wings
came up. They were pulling a cart with a huge tub of water
on it. They writhed along, snapping their tails proudly like a
bullock-driver's whip. All round the cart swarmed men,
ladling the water from the tub and sprinkling it on the clouds,
like gardeners watering the soil.

When the men caught sight of Yo, they stopped what they
were doing. 'Who are you?' they said. 'Cloud-stranger! What
do you want?'

Before Yo could answer, one of them stepped up and took
his hand. It was the giant stranger from down below on
earth.

'It's all right,' he said. 'This man is my friend. Give him a
ladle, and let him help.'

So the men gave Yo a ladle. He began to water the clouds,
and the dragon-cart went on its way.

As they went, Yo's friend explained that they were servants of the Thunder-god. Their job was to distribute rain
evenly across the earth. It was because of a mistake he had
made in this, three years before, that he'd been condemned to
walk the earth until he found an honest man.

'You helped me, Yo,' he said, 'and now you'll have your
reward. Down below are your own two fields. Sprinkle
hard!'

Yo filled his ladle and splashed as much water as he could
on the fleecy clouds. Then the Thunder-god's servants let

him down from heaven on a long rope twisted from the dragon-reins. He landed with a bump, just outside his village, and the rope was drawn up again out of sight.

Yo walked down the road to his house. All round, in the dry sun, the fields of his neighbours were withered and parched. But his own fields were fresh and green; water bubbled and glittered in the water-courses; his crops stood tall and strong.

When Yo reached home, he took the star out of his sleeve and put it on the table. It was dull and brown, like an ordinary pebble. But at night it began to glow, and its brilliance filled the house with light. It became Yo's greatest treasure. He kept it carefully, and only brought it out when he had important guests, to light them as they ate.

In China long ago, people believed in the existence of spirits and demons, and thought that dreams had the power to foretell the future. There are many stories and legends of warnings sent to wicked men; this one is rather like a Chinese version of Charles Dickens' A Christmas Carol.

Tseng and the Holy Man

Tseng was a handsome young man, well educated and talented. But he was also bumptious and boastful, as full of himself as a soap bubble. Everyone admired him, but nobody liked him. People smiled at him and rubbed their hands with pretended pleasure when they saw him – but only because they were afraid of what he might do when he became rich and powerful.

Rich and powerful! Tseng's ambition burned in his brain. The three things he wanted more than all the world were the dragon robes, red umbrella and brass-bound treasure-chest of a Minister of State.

One day Tseng and a group of other young men, out walking in the countryside, were caught in a storm. Raindrops the size of duck-eggs pelted down; their sandals slithered and squelched on what had once been a dusty track; their clothes hung dark with rain. Even Tseng was soaked, for all his airs.

It was a case of finding shelter – anywhere, and soon. They didn't want to drown on dry land. They squelched through grey rain to the only house in sight: a broken-down hut of mud and straw. Its walls ran with water and wind roared in the eaves, but at least it had a roof.

Tseng pushed open the door and shouldered his way inside. The others crowded in after him. The hut was small and dark. A fire of dried dung flickered and spat. An old holy man sat cross-legged on the ground, still as a lizard. He took no notice of his guests. They fed and poked the fire, squeezed the wet from their clothes and sat down to warm the storm from their bones.

Soon, in the dark hut, the murmur of voices mingled with the thud of rain on the roof and the hiss of the fire. Tseng was drowsy, lost in a daydream of what he would do when he was rich and powerful. He had just, in his imagination, made his old gardener Prime Minister, when there was a jingle of harness from outside, a knock on the door, and in walked two heralds, dressed in the royal robes of the Emperor himself. One of them carried a red velvet cushion with a yellow scroll upon it; the other held a small umbrella to protect it from the rain.

The heralds bowed low to Tseng and handed him the scroll. It was nothing less than a letter from the Emperor's own secretary, inviting him to Peking to become a Minister of State. The heralds dressed him in dragon robes, gave him a red umbrella to keep off the rain, and led him to a fine house in the city, with a brass-bound treasure-chest in every room.

★   ★   ★

Tseng's job was an important one. He was an Official of the Second Grade, and was responsible for the promotion or dismissal of all the Officials below him, from the Third Grade down to the Ninth. He was a powerful man, and everyone wanted his favour. Presents of costly food arrived daily in his kitchens; his hall was thronged with flatterers; his treasure-chests bulged with gifts of coins, jewels and embroidered silk.

When Officials of the First or Second Grade came calling, Tseng would dress quickly in his finest robes and rush out to fawn and wheedle them, rubbing his hands and smiling his smile. But with underlings, Officials of the Third Grade and below, he was hard, cold as stone. One nod and a man was stripped of his power; one finger crooked and another man was thrashed without mercy. Power – and wealth! In a few years Tseng was richer than dreams, almost as powerful as the Emperor himself.

But power and wealth breed envy – especially if they

grow from arrogance. As the years passed, other Ministers envied Tseng more and more; their envy curdled to hatred and their hatred into spite. Finally they sent a petition to the Emperor: all those who had once thronged to Tseng's door, who had fawned on him and called him 'Little Father', now accused him of treachery, cruelty and feeding fat on bribes.

The Emperor's answer was swift. The same two heralds (older and slower now) came to visit Tseng again. They brought another velvet cushion, another yellow scroll. Banishment, for life. They were followed by a whole army of removal-men, with mule-carts. All Tseng's treasure-chests, his wardrobes, his vessels and bales and crates, were stacked in the hall. Carts were piled high with bank-notes, jewels, silks, fabrics, caged humming-birds, wine-barrels and honey-jars. Curtains, carpets and bedspreads were folded and baled; a thousand pairs of shoes were wrapped in silk; lampshades, silver tea-caddies, ornaments of jet and jade, were crated and packed. The laden carts trundled away; the dust settled again on the cobble-stones; a detachment of squint-eyed, broken-down soldiers marched up in ragged step, to lead Tseng and his wife into exile, a thousand leagues away.

A thousand leagues! On foot, mile after endless mile. Mountains; marshes; deserts; stony tracks. Getting to exile was a lifetime's job. Or it would have been, except that on the first night of the second month, just as Tseng and his wife were settling down on a bed of desert stones, a band of robbers came down on them: leather breeches, shouts, moustaches and bristling swords. What use were the soldiers? None: fast as ferrets, they ran away.

'Spare me! Spare me!' cried Tseng, on his knees. 'I'm a poor man, an exile. Take my wife; spare me!'

'No!' thundered the robber chief. 'Look at our faces. We're your victims, the men you stripped and drove into exile. Now we want your head.'

'Dogs!' said Tseng. 'Exiled I may be, but I am still an Official of the Second Grade, your superior. Bow and fawn!'

But the robbers crowded round, and the next thing Tseng heard was the plop of his own head as it hit the ground. Darkness loomed around him. And out of the velvet dark

two blood-red devils appeared, chirping like monkeys. They
took Tseng's hands in their scratchy paws and dragged him
after them, chittering and squeaking with delight. As they
went, a smooth stone path, blue as iron, opened up in front of
them: the way to Hell.

★   ★   ★

They came to the city of the Emperor of Hell, and the
hideous palace where he sits in state to judge the Dead. The
devils flung Tseng on hands and knees and he crawled before
the Emperor, whimpering for mercy. The Emperor turned
to one of his Officials, who held a long grey scroll: the
register of mankind and their deeds on earth.

'Name?' said the Emperor.

'Tseng,' answered the Official.

'His first crime?'

'He betrayed his friends.'

'The punishment: boiling oil.'

The Emperor gestured with his hollow hand, and all the
ten thousand devils gathered in the hall shrieked with glee.
They led Tseng to a brass cauldron two metres high and two
across, filled with bubbling, smoking oil.

Tseng howled for mercy, but no one heard. Chuckling,
the devils picked him up and dropped him into the oil. He
splashed, sank and bobbed up again, like a fish deep-fried on
the kitchen stove. He felt the burning oil seep into his body; it
gnawed his bones till he screamed for death. But he was dead
already, and all he could do was writhe and twist till the
punishment was done. Then a devil fished him out with a
boat-hook and dumped him down in front of the Emperor,
floundering, gasping, sodden with oil.

'Tseng!' said the Emperor. 'Now for your second crime:
injustice and insult. You used your position to harm others.
Punishment: the hill of knives.'

The hill of knives! Compared to it, boiling oil was a
pleasure, a soft delight. The devils took Tseng to a hillside
studded with knives, set point-up in the ground like hedge-
hog's bristles. Bones and rags of flesh: the remains of
previous victims. The devils tossed Tseng high in the air and

let him fall on the points. Again. And again. Then they gathered the pieces in a basket and dumped him before the Emperor once more. Tseng groaned; he tried to hold his tattered flesh together; he begged for mercy.

'The third crime?'

'Embezzlement. He stole three million gold pieces.'

'Punishment?'

'He must drink the same amount.'

Three million gold pieces, melted down – a river, a searing waterfall. The devils thrust a funnel down Tseng's throat and began pouring in the molten gold. Gulping, twisting in agony, he drank and drank. When it was done and his punishments were over, the devils took him to a catherine wheel the size of a man. They fastened him to it and lit the fuse. Round the wheel fizzed, faster and faster. Light and darkness blurred together, burning Tseng's eyes. He screamed and screamed.

★ ★ ★

Well, it was a dream. Of course it was a dream. Tseng woke to find his companions shaking him, trying to snatch him from his nightmare. He sat up and shook his head. The holy man's lizard-eyes were open, and a smile flickered on his papery lips. 'Tseng, Tseng,' he said in a voice like bank-notes blown on a summer breeze, 'Have you learned? Do you know your future now?'

Tseng bowed his head. Outside, the rain was finished; the sun shone; crickets chirped; the green ground steamed. He went home an altered man. Ambition is not an evil thing; but evil ambition always destroys a man. Be honest and true: for good men, even in the volcano's heart, a lily will always grow. So Tseng was saved from ambition; he lived a long, uneventful life, honest and true, loved by all his friends; and when he died the devils carried him to happiness, the true reward for loyalty.

And the holy man? He waited, cross-legged in his hut like a lizard in a wall, to set the next visitor's feet on the path of truth.

The civilization of ancient Greece gave to the world a rich treasury of myths and legends. This story tells how Prometheus, an earth-god, created man; and how Zeus, king of the gods, took his revenge for a trick Prometheus played on him. It is really about knowledge. If man has knowledge, he knows about evil as well as good. If he chooses, then, to do evil, how can god be pleased? Is it not better for men to be ignorant, their minds as empty, and as pure, as animals'?

Prometheus

CLAY–MEN

At the beginning of time there were no men, no women. Only a great darkness and a great light. The darkness was the deep, purple velvet of space. The light glinted and glowed from the marble palaces of Olympos, home of the gods. They galloped their chariots down the corridors of space, along the wide road of the Milky Way; in Olympos they took their places on polished thrones, at the councils of the gods.

The gods had a beautiful garden to walk in, called the world. Quiet orchards, rippling meadows, placid lakes – or, if they chose, seas snarling at jagged cliffs, barren snowfields, glacial peaks in the thin raw air. The world was their pleasure-ground, the plants their joy, the wild animals their pets. There were no men: men would have asked for a share in the garden, a corner of the world for themselves.

Only one god, an earth-god more ancient than time itself, took no pleasure in the emptiness of the world. His name was Prometheus: Forethought. He was a Titan, huge and craggy, a living hill. He tramped the echoing world, looking out at its emptiness. Sadness filled his heart like mist. So much beauty . . . such peace . . . and all for the gods alone.

Then, out of nowhere, an idea flickered . . . grew stronger . . . shone white and clear. The Titan smiled to himself, and squatted down on a bank of brown, oozing clay.

His thick hands gathered it and he began working it slowly, clumsily, in his stubby fingers. First a body . . . then arms . . . legs . . . a head. A figure copied from the shape of the gods. One . . . then another . . . then another. He worked on, heedless of the world's seasons changing around him, till a pile of naked brown clay-men lay ready at his feet.

Prometheus sat back and looked with pleasure at his creation. Model people . . . a doll-family to fill the empty world.

There was a sudden sound and he scrambled awkwardly to his feet. A whirr of wings, a glitter of light . . . There, settling on a rock, was a seabird. A seabird with eyes dark and fathomless as mountain lakes. It was one of the gods. Prometheus flushed and tried to scuff his clay-men out of sight.

'What are they, brother? What magic have you made?'

It was Athene's voice. Prometheus picked up one of his men and showed it to her, naked as a baby on his huge, hard palm. Athene bent her head and blew softly into the clay-man's nostrils. Prometheus felt her warm breath dance on his fingers, like sunshine.

Then suddenly there was a tingling, a wriggling, a twisting of life in his hand. The clay-man sat up, stretched his arms, yawned – slowly and awkwardly, as if he was learning the movements even as he made them.

One by one Athene gave the clay-people the breath of life. They stood up, unsteady as newborn foals, and tottered about uttering tiny cries. They tumbled and squealed round the Titan's feet, like kittens fumbling for their mother.

Mankind was born.

When the gods heard, they were furious. The storm of their rage flickered and crashed across the universe, darkening the stars. Prometheus was called before Zeus the Thunderer, king of gods. The king's anger, like a thunderstorm, a hurricane, crashed about his head, till even the Titan flinched.

But nothing could be done. What was made could not be unmade, even by Zeus. At last the king crashed his sceptre on the marble floor and said, 'Very well! Your men must live. But they will live innocent and free. No pain, no guilt, no memory. They will know nothing, understand nothing. They will be sinless, like the beasts in the fields.' Prometheus bowed his head. 'Yes, Lord Zeus,' he said. 'But are men to know the gods? Will they know we exist?'

'They will know us and fear us,' said Zeus. 'They will worship us, and sacrifice the best of their flocks and herds to us. See to it!'

Prometheus went down into the world. His men were scurrying about, busy as ants. He watched them, and pitied them. 'They are so clever, so like the gods,' he thought. 'Why must they be sinless? Why must they know nothing but fear?'

From Olympos, Zeus looked down. He saw the toiling men, and Prometheus like a hillside above them. He saw the pity in Prometheus' mind. The father of gods acted. He sent for Athene and for Hephaistos, craftsman of the gods. 'Hephaistos,' he said, 'fetch clay, and make me . . . a woman. Model her on Aphrodite, the most beautiful of all goddesses. Athene will give her the breath of life. We'll call her Pandora: All-gifts, and hide her safe till we need her. Go!'

So plans were made in Olympos. In the world, Prometheus made plans as well. He took a magnificent bull, the best, most perfect animal in all the world, and sacrificed it. He cut up the meat. All the bones and fat he bundled up, and wrapped carefully in the beautiful, creamy-brown hide. The good, lean meat he stuffed in a bag of guts. Then he shouldered the bundles and went back to Zeus.

'Choose, Lord,' he said, 'one bag for the gods, the other for men. The way you choose will fix the way of sacrifice forever.'

Zeus looked at the bundles; the smooth, unblemished sack of hide and the oozing bag of guts. 'The gods' portion is . . . *this* one,' he said, pointing to the hide. But when Prometheus opened it and showed a pile of bones and yellow fat, his anger swelled and roared, till marble Olympos shook.

Like a meteorite, a shooting star, Prometheus was hurled back to earth. He crashed down, the bag of guts clutched in one hand. Zeus had chosen, and the choice would last forever. But mankind, Prometheus' creation, would suffer for it.

Up in Olympos, Zeus sent for Pandora. 'Go down,' he ordered. 'Take men this box. It must never be opened. Keep it safe, and go.'

Pandora fell to earth, holding the box. It was a small, wooden box with leather hinges and a bronze lock. She fingered it, turned it over and over, wondering what could be inside, what secret was to stay locked away forever.

One by one, like inquisitive forest creatures, men stole up to see her: the first woman. Their eyes were bright with curiosity, but no understanding. They were like beasts: they knew things by instinct, but had no power of thought.

Curiosity ached in Pandora. Her fingers strayed over the hinges, the lock . . . The men watched, placid, like deer in a park.

At last she could bear it no more. She fumbled the lock apart and slid the box open. A low buzzing began, and grew and grew and grew into a cloud of grey insects that flew up and blanketed the sun. Pandora screamed and screamed. The men fled. Pandora dropped the box and ran.

Alone, Prometheus bent for the box. He was slow and sad. The insects were Zeus' gift to men: plague, pestilence, disease . . . all the torments of the flesh. Prometheus' trick was paid back, his defiance punished. Mankind had been given pain.

Tears furrowed his cheeks. He let the box fall to his side. There was a tiny rattle from inside. He lifted it again and peered into the corner. There, lost in the shadows, smaller than the smallest speck of grain, was a tiny, shrivelled seed.

Carefully, so as not to crush it or drop it, Prometheus locked it back in the box and hid it safely away. Zeus had shown mercy after all. His gifts were not a destruction. Among all the plagues and torments sent to men, he had planted a single, tiny seed . . . of hope.

FIRE-THIEF

Down on earth, the first snow was falling. Softly, silently, it whispered across forests and fields and hills. It settled gently on the branches like a soft, white death. It slipped soundlessly into cracks between the rocks, and choked the gullies where sheep huddled and died. In gaunt lakes and the still, grey sea snowflakes disappeared and died. There was no sound.

Here and there, in caves in the rocks or under trees, men gathered. They were naked, their skin putty-coloured and pitted with cold. They cowered together in groups for warmth. With gentle, mindless eyes they looked out at the falling snow. They saw it, but could not understand or wonder at it. Without intelligence, they saw no link between the white drifts and the cold that dulled their bones. Like sheep, they huddled and watched, and died.

Up above, in the palace halls of Olympos, the gods were feasting. Iron fire-baskets glowed on the dusty floor and threw flickering shadows on the walls. The pillars gleamed and shone in the warmth, and the gods lolled at their ease, laughing and talking without a single care.

Only Prometheus sat silent. He looked into the heart of the fire and shivered. His heart burned with pity for mankind in the world below. His creation, chilled by their own ignorance, withering with cold. If they could once understand, if once the spark of knowledge glowed, their minds would thaw and they would learn the secrets of nature, the secrets of the gods.

But knowledge, fire, was not his to give. Any god could create, breathe life into lifeless things; but intelligence was allowed by Zeus alone.

And Zeus had been tricked. He had planned to create a perfect, sinless race. But Prometheus had tricked him and blocked his plans. Zeus could not uncreate what another god had made. But he could stand by, do nothing, while nature ran its course. The cold would finish men. Mindless and ignorant, they would perish soon enough. All Zeus had to do was forbid them fire.

Prometheus stood up. Wrapped in his dark cloak, he blended with the shadows. No one noticed as he went out. Sweating, laughing, joking, the gods continued with their feast.

Wrapped in his cloak, soaring like a nightbird, Prometheus passed from the home of gods to the home of men. He made his way deep under the earth, to the caverns of Sicily. Here, in the searing heat below the earth's crust, Hephaistos the blacksmith-god ruled. His servants, the brutish Cyclopes,

sweated and toiled, shaping the iron bones of the earth into thunderbolts for Zeus himself. The fires flickered and roared; the hammers thudded and the anvils groaned; waste rock, white-hot, flowed out through vent-holes on the surface, the volcanoes of Sicily.

Unseen, Prometheus knelt by a stream of glowing fire. From his belt he took a stalk of green fennel. He split it open. The inside was hollow and damp. It would keep a glowing coal alive for the time he needed it. Quickly he took a coal, slipped it inside the stalk and drew back into the shadows.

There was a jar of silence, a single moment's pause in the lava-flow. The Cyclopes looked up. Something was gone: a fragment of the god's fire was missing. Hephaistos heaved himself up on his twisted, withered legs and limped across the cavern.

But there was nothing. No one was there. The work began again.

Outside, on earth's surface, falling snowflakes sizzled against Prometheus' seared, charred skin. His lungs were scorched, his flesh burned raw, his strength withered. But in his belt, in the fennel-stalk, was a fragment of Zeus's all-powerful fire. He bent, and blew on it till the coal glowed alive and red.

Fire! Prometheus' last gift to men. The fire of intelligence, to save them from death and teach them the gods' secrets.

Years, generations later, the time came for Zeus to survey the world, to check that cold and storm had rid the gods' garden of Prometheus' doll-creation forever.

But the world had changed. From every part of it, pencils of smoke rose up from the wink and glint of cooking-fires. Men hurried about, eager and purposeful. They wore clothes now, animal skins and woven cloth. They had learned farming, sailing, a thousand crafts and skills. They were busy with houses and markets and parliaments. They were fishing, hunting, harvesting. Above all, they had learned speech: a hum of voices, a bee-swarm of languages, rose up and drowned the murmur of the sea, the rustle of the wind.

Zeus' anger was a lightning-storm, a rolling cloud, a rumble of thunder. The sky darkened; the earth shook; the gods flinched.

For Prometheus, Fire-thief, there was no escape. There was the hiss of a thunderbolt as it found him out, a flare of light, an eruption of agony that filled his body and ripped his mind. Black, greasy smoke; a stench of charred flesh; silence.

There was more to come. For a fallen angel, an immortal, punishment is immortal too. Might and Force, servants of the gods, gathered up his tattered body. Hermes the Messenger led them, and Hephaistos limped beside them with chains of unbreakable steel. Prometheus' head lolled and gasped like a gaffed fish, as he was carried to the place they chose for him.

On a high mountain-peak, a jagged tooth of rock stands gaunt and black. Cloud coils round it, dank as corpse-breath. No trees; no animals; no life. Only a needle of stone in the thin, cold air.

This rock was Prometheus' place. Here Might and Force stretched him on tiptoe, and held his arms high above his head while Hephaistos chained him to the rock. Unbreakable steel. Forever.

Their job done, the gods went back to the warmth of Olympos. Prometheus was left alone. He could see nothing, hear nothing. His charred skin stuck to the icy rock; his muscles locked hard with cramp; his wrists and ankles were raw where he writhed against the chains.

Then suddenly, in the echoing silence, he heard a whirr of wings. A black shape loomed over him, dark in the cloud. There was a searing, stabbing pain as a curved beak tore into his ribs, stripping its way past bone and sinew till it found his liver.

The vulture of Zeus. All day it gouged and tore his flesh. At night it rested, and ragged wounds healed ready for the next day's pain. Prometheus, Fire-thief, was locked in an eternity of suffering. He knew before he stole that this would be his punishment. He knew it, and chose it. To save his own creation, he chose to destroy himself.

MOTHER'S BONES

So Prometheus suffered the punishment he had brought upon himself. For an age of the world, while a thousand

generations were born and died, he endured his torment. The
anger of Zeus simmered like a distant lava-flow.

But to be all-powerful, all-angry, is to be all-merciful too.
Alone of the gods, Zeus knew justice and forgiveness; alone
of the gods he had free will, the power of choice. The time
came when he decided that Prometheus had suffered enough.
Heracles was sent, with unerring arrows; the vulture was
killed and Prometheus' chains unbound. Blessed by Zeus's
mercy, he lived on in the halls of Olympos, a huge, forgotten
power from a former time. His age was past; a new age of
justice and mercy had begun.

First justice, then mercy. There was still the problem of
Prometheus' men. In the centuries of his punishment they
had spawned and spread, till they covered the surface of the
earth. The curses from Pandora's box had settled in their
hearts and filled them with evil and crime. Hope was forgotten, a seed no one bothered to plant.

Above all, mankind forgot the gods. If they remembered
Zeus, Prometheus and the great powers, it was as legends,
dim figures in children's story-books. Pride ruled; men
thought there was nothing they could not do. What need had
they of gods? Swollen with pride, they grew fat on insolence
and crime. To the gods looking down, the world was like a
carcase, swarming with maggots.

Zeus sent for Prometheus. Warily, the Titan bowed his
head. He already knew the thoughts in Zeus's mind.

'Your men, Prometheus; your creation. The whole world
is polluted. It is time to wash it clean.'

'They must all die?'

'No. Even now, there's a little good left in them. Choose
one man and one woman, and tell them a way to escape the
flood. The rest will drown. When the world is washed
clean, your couple will begin a new, sinless race that fears the
gods.'

At first the rain was gentle and soft, like a summer shower.
People splashed through the puddles, laughing and shaking
the drops from their eyes. It was a freak shower; it would
soon pass.

But instead of easing, the rain grew heavier. Day after day,
cold winds gusted in and whipped the drops to torrents.

Puddles turned to ponds, ponds to lakes, lakes to seas. Rivers burst their banks and flooded the countryside. Crops were ruined, animals drowned, houses and villages engulfed.

As the waters rose, the people crowded for safety on to higher and higher ground. Now they remembered the gods: remembered and cursed them. Prometheus' creatures were as they had been before he gave them fire: knowing nothing, understanding nothing, they huddled together against the storm, and died.

Only two people knew what to do. Their names were Deucalion and Pyrrha, his wife. Quietly, carefully, they built a floating shelter, a wooden ark that would rise higher than the highest flood, and save them from death.

Finally, Zeus gave the word to his brother Poseidon, god of the sea. Poseidon opened his floodgates, and the water roared out across the world and swallowed it whole. People and animals were swept away like toys; houses splintered; for a time there was crashing, rumbling and shrieking – but not for long.

The tidal wave filled every gulf and cranny of the world, till there was nothing but smooth, calm water. Bobbing on the surface, like a speck of dust on a pane of glass, was Deucalion's ark.

Bubbling and gurgling, the water began to recede. First mountain-tops, then forests appeared, then muddy lowlands steaming in the sun. There was no life: as the water slid away it carried with it every trace of living things. The world was as empty, as fresh and clean, as at the first creation.

Deucalion and Pyrrha beached their ark on a hillside, Mount Parnassus in Greece. They stepped out on to the damp turf and stared round in dismay. No birds, no insects, no animals. Nothing. The world was still, except for water dripping and gurgling on every side.

How were they to live? They were alone and helpless, in an empty world. They looked at the sun dappling the forest leaves, and shuddered.

Then, all round them, the whisper of a voice came and filled the air. Like leaves rustling in the wind, it shaped itself into syllables and groups of sounds, words that seemed to make no sense.

'Mother's . . . bones . . . gather . . . throw . . . Mother's bones . . . mother's . . . bones . . ."

Deucalion and Pyrrha looked at each other. Their eyes bleached with terror. How could they gather their mother's bones, now, after this flood? And where were they to throw them? *Why* should they throw them? Were the gods mocking them, with such an insane command?

At last the answer formed in their minds. Their mother meant Mother Earth, and her bones were stones. Deucalion picked up a pebble and threw it carefully over his shoulder, holding his breath to see what would happen.

As soon as the pebble touched the ground, it turned into a young man: tall, laughing, strong.

Slowly at first, then faster and faster, Deucalion and Pyrrha began gathering pebbles and throwing them over their shoulders. Deucalion's stones turned into men and Pyrrha's into women. The new people stood in a still group, watching their creators as they ran down the hillside tossing pebbles behind them as they went.

New men. Prometheus' creation, reborn. Was this Zeus's sinless race at last?

Gilgamesh was king of the city of Uruk in Sumeria over four thousand years ago. The Epic of Gilgamesh contains a series of legends associated with this shadowy figure, said to be two-thirds god and one-third human. The story of Utnapishtim and the flood is part of the legend of Gilgamesh and his quest for immortality.

Utnapishtim and the Flood

Gilgamesh was mourning the death of Enkidu, his greatest friend. The sight of Enkidu's face in death had made Gilgamesh realize that he, too, must die one day; now he wanted to find a way of living forever. He remembered that he had heard stories, long ago, of an old, old man named Utnapishtim: he lived on a distant island, far away at the end of the earth, and he knew the secret of immortality – he was the only human being who would live forever. Gilgamesh determined that he would journey to the island, find Utnapishtim and ask him the secret.

The journey was long and hard. Gilgamesh travelled over sun-baked, trackless deserts, past gaunt mountains rigid with snow, through cities and continents sprawling with human-kind. He came to the sea that laps the edge of the world. The waters of death! Except for Utnapishtim himself, no living man had ever crossed them. But now Gilgamesh stood over Urshanabi, ferryman of the dead, and forced him to give him passage in his leaking, rotting boat. They rowed out across the barren sea, day after day, surrounded by emptiness.

After many days the boat came in sight of land: an island, like a single speck of sand in the endless sea. As they came nearer, Gilgamesh made out a figure sitting watching on the shore. It was a man: the oldest man he had ever seen. His body was bent and shrunken, like a puppet's when the strings are slack; his brown, wrinkled flesh lay like paper on his bones; his thin white hair was wispy on his head. With

fathomless, sunken eyes he watched Gilgamesh land. The ferryman paddled out of hearing, to wait.

Utnapishtim! The only mortal to be given immortality. Gilgamesh's search was done.

At first, when he heard why Gilgamesh had come, Utnapishtim drew back in fear. 'The secret of everlasting life!' he said. His voice was a thin whisper, as if he had almost forgotten how to speak, long years ago. 'How can I tell you that? All men must die. Lord Marduk, who created man, made it a condition of our birth. You must die, so that your life can pass to others, for them to enjoy. Your life was lent to you by the gods; it was not a gift.'

'I know that,' answered Gilgamesh. 'Did I not see life depart from Enkidu, my companion, the friend I loved? But what of you, Utnapishtim? What of you? You are a mortal man: you have eyes, lungs, feet, a heart, like every other man. How is it that you alone have been given your life to keep? And if you succeeded, why can't I?'

Utnapishtim sighed, like wind gusting through distant reeds. He would have to speak. Only when Gilgamesh heard his story, would he realize that the secret of life was not his to tell. No other mortal man could ever follow the path Lord Marduk had given him.

'A thousand generations ago,' he said, 'when I was young, the gods began to be troubled about the men Lord Marduk had made. It seemed to them that there were far too many men on earth. They were increasing all the time. There was not enough work, not enough food; they were growing noisy and quarrelsome.

'At last the gods decided that the only way to deal with men was to flood the whole world, and let them drown. Just before the flood began, Ea the water-god warned me of the disaster to come, and ordered me to build a boat. Under his instructions, I built a wooden boat big enough to carry my wife and family and all our possessions, together with the workmen who had helped with the building, and all our animals. We caulked the seams with tar, and stocked the boat with food and water. When the other inhabitants of our city asked what we were doing, I answered that the water-god Ea had ordered me to leave the land and live on water. This he told me to say; I was not allowed to warn the others.

'The storm broke; the flood waters rose. For six days we rode out the storm, while all the earth was covered and all other men perished. The fury of the flood was so great that even the gods were terrified, and cowered in the corners of heaven like dogs.

'On the seventh day the storm died down. I lifted a hatch and looked out. All I could see was water: flat, motionless, deserted. There was no coastline: the earth had been swallowed up by the sea. I wept for my city and my fellow-citizens.

'Then the waters began to subside, and we grounded on Mount Nisir. After seven more days I sent forth a dove; but she found no other land, and returned to us. Next I sent a swallow, and she too returned to the boat. Finally I sent out a raven, and the raven found land and food and did not return. We were able to release the animals, and made a sacrifice on the mountain-top to thank the gods for our safety.

'Enlil, the storm-god, was furious with Ea for giving away the secret of his flood and allowing us to escape. But Ea calmed him down. Then, because I had exactly carried out his orders, he gave me a reward: the gift of everlasting life.

'This gift, which you crave so much, is double-edged. True, I shall live forever. But because the gift may be shared with no other mortals, I am banished to this remote island forever. Everlasting life – and unending loneliness.'

So Gilgamesh at last realized that Utnapishtim had no secret to pass on to him. His immortality was a gift from the gods, and there was no way to persuade the gods to make such a gift again. Mortal men are not meant to live forever. We have a short time to live on earth, and while the time is ours, the secret of happiness is learning to enjoy it and make the most of it.

The best-known version of the universal story of God sending a flood to punish mankind is that of Noah, in chapters 6–8 of the book of Genesis, the first book of the Bible. Whether or not a flood which covered the whole earth is scientifically possible is perhaps irrelevant; ancient man no doubt thought of the valley in which he lived as the whole world, and there is plenty of evidence that the Tigris and Euphrates valleys in Mesopotamia, from which area this story probably originates, suffered devastating and unpredictable floods.

Noah's Ship

Many years after Adam and Eve had been banished from the Garden of Eden, their descendants had spread out all over the land and were indulging in all kinds of wickedness. Things got so bad that Lord Yahweh was sorry that he had ever made man in the first place, and made up his mind to wipe out the whole of creation – men, women, animals, birds, everything.

There was only one person Yahweh thought worth keeping; this was Noah, a good, kind man who respected Yahweh and drew back from the evil and corruption around him.

Yahweh took Noah on one side. 'I am sickened by the behaviour of mankind,' he said. 'You are the only one whose behaviour pleases me. This is what I mean to do. Soon, I shall make it rain so hard, and for so long, that the whole world will be flooded. You must build a ship, on dry land, big enough for all your family and some animals and birds – seven pairs of the kind you eat, and just one pair of each of the others. When the flood comes, everyone else will be drowned, but you will be safe in the ship. Be sure to take enough food with you for yourselves and the animals – the flood will last for quite a time.'

Noah did exactly what Yahweh had told him, and then waited for the rain. One week after the ship was finished, it started to rain. Noah, his wife, and his three sons went aboard the ship with the animals and birds, and Yahweh shut the door behind them.

Before long, the flood waters began to rise and the ship with its cargo began to float. The flood waters spread until every inch of land was covered, and everything living – except Noah and his shipful of people and animals – was drowned.

The rains lasted for a very long time. Noah's stocks of food were running low and he began to wonder how they would all survive. Then, at last, when Yahweh was sure that everyone else had drowned, he made the rain stop and the flood-waters recede. Noah's ship continued to float until, finally, it grounded on what turned out to be the top of Mount Ararat.

Soon the tips of other mountains could be seen, and Noah opened a hatch and let a raven fly out to look for land. But the raven kept flying to and fro over the ark, without finding anywhere to settle. Noah decided to wait for the waters to go down some more.

A week later, he opened the hatch again, and this time sent out a dove. The dove flew off, but still there was no dry land or trees for her to perch on. The dove flew back to Noah, who stretched out his hand and took her back into the ship.

After another week, Noah sent out the same dove once more. This time, she was gone all day, and flew back late in the evening with an olive leaf in her beak. The olive leaf was shiny and new, and Noah realized that the flood level must be down below the treeline. A few days later, he sent the dove out again, and this time she didn't come back at all.

Noah knew now that it was safe to leave the ship. He took off the hatches and let all the animals and birds out onto the mountain side. He kept back some of the animals, built an altar of stone on the mountain top, and sacrificed as a thank-offering to Yahweh for their safety.

Yahweh smelt the sacrifice from far off, and thought kindly of Noah and his family.

'I shall never again drown with floodwater the living things I have created,' he said to himself. 'Even though man's very nature seems to make him do wrong, I will not do that again. As long as the earth lasts, I will let the seasons of seedtime and harvest, winter and summer, sunshine and rain, continue without interruption.'

*Once again, mankind is threatened by a great flood. This version of
the ancient story comes from India, and is put together from several
versions in the Hindu Scriptures.*

Manu and the Fish

Manu was a wise and holy man who devoted his life
to the worship of the gods. One day he was praying
by the banks of a river, when he heard a tiny voice
calling.

'O great and holy Manu,' the voice said, 'I appeal to you
for help. You're a holy man; it's your duty to help and protect
the weak.'

Manu looked round to see who was speaking. But there
was no one there. Then the voice came again. 'In here,' it
said. Astonished, Manu saw that it was coming from a tiny
fish in the river in front of him.

Carefully, he filled his cupped hands with water, bent and
brought out the fish.

'Tell me what I can do for you, little one,' he said.

'Help me,' said the fish. 'You see how small I am. In the
river I'm the smallest living thing: my life is in danger from
enemies of every size.'

'What must I do to help?' asked Manu.

'Put me in a jar, safe on land,' said the fish. 'In return, I'll
save your life as well. A great flood is coming; all mankind
will be swept away. If you help me, you alone will escape.'

Manu did not take this promise very seriously: even if
there was a flood, how could a tiny fish save him? Never-
theless, he fetched a large clay jar, filled it with water and
placed the little fish in it.

Now this was no ordinary fish. In no time it had grown too

big for the jar, and Manu had to dig a pond for it outside. Before long, even the pond was too small, and Manu carried the fish, which was now huge, to the river Ganges.

Before it flopped into the river and made off to the ocean, the fish spoke again to Manu. It told him when the flood would come, and gave him instructions. He was to build a large boat, in good time, so that he would be safe when all the land was flooded. In it, he was to put seeds of every kind of plant, and a length of strong rope. The fish promised that he himself would reappear, and save Manu when the floods came. Manu would recognize him by a large horn on his head.

As the fish swam away, Manu realized that it was not really a fish at all, but the god Vishnu, preserver of life, himself. No mortal could look at a god and live; so Vishnu had appeared to him as a fish, in a form he could recognize and understand.

Respectfully, Manu bowed his head; when he looked up, the fish had gone. Manu hurried home and began to build a boat, as he had been instructed. He gathered seeds of every kind of plant, and plaited a length of strong rope. When all was done, he waited for the flood.

When the storms began and the waters rose, Manu set out in his boat across the sea. Before long, the great fish reappeared, and Manu made a noose in his rope and fastened it to the horn on the fish's head. Towed by the fish, Manu's boat was safe from all danger.

At last the waters began to subside. The boat grounded on a mountain-peak high in the Himalayas, and the fish ordered Manu to moor it to the rock. Manu was the only survivor of the flood that destroyed mankind. He was the first of a new race, the father of all.

Another story of God seeking to punish mankind by a flood, this time from China.

The Yellow Emperor and the Great Flood

High in the heavens, the Yellow Emperor sat musing on the wicked ways of mankind. Everywhere he looked on earth, he saw nothing but evil. Nothing, he decided, would bring man to his senses. Unless . . . a flood? Thought led at once to action. The Emperor of Heaven sent for Kung-kung, Spirit of Water, and gave him his instructions.

Kung-kung was a cruel god who took delight in using the power of water to cause suffering. Chuckling, he raced across the skies, causing torrential rain to fall on earth. For many days it rained, until the rivers burst their banks, and the floodwaters swirled through every town and village. Fearfully, men and women gathered up their belongings and made for higher ground. Many were drowned as the floodwaters overtook them, and those who escaped were soon starving for want of food.

Only one of the gods was moved to pity by the plight of mankind; Kun, grandson of the Yellow Emperor. He went to his grandfather and pleaded with him to recall the Spirit of Water and allow the floodwaters to abate. But the Yellow Emperor would not relent.

Sadly, Kun walked away from the imperial court, wondering what he could do to help mankind. As he walked, a gruff voice broke in on his thoughts.

'Why are you so miserable, Kun?' the voice enquired. It was a black tortoise, one of the oldest and wisest of creatures.

'I grieve for the suffering of mankind,' replied Kun. 'I

would dearly like to save the world from the flood, but I know not how.'

The tortoise snorted. 'No difficulty about that,' he said, in his deep, slow voice. 'All you need is a handful of Magic Mould.'

'Magic Mould?' exclaimed Kun, not sure whether to take this strange creature seriously.

'Magic Mould,' replied the tortoise. 'It's a special sort of earth; sprinkle it on water and it grows and expands to any amount you want. You'll only need a little.'

'But where would I get it from?'

'From your grandfather, the Emperor, of course. Who else?'

Kun sighed. 'That's no good, then,' he said. 'He would never give me any.'

'So, steal some!'

Kun sat down, and the two of them worked out a plan for stealing a handful of the Magic Mould. No-one knows quite how Kun did it; but of course, being a god, he could change himself into many different shapes to deceive the guards and get into the room where the Magic Mould was kept.

When at last he had a supply of the Magic Mould, Kun set off to earth to try it out. Sitting on top of a high mountain, he sprinkled a few grains of the Mould onto the surface of the floodwaters beneath his feet. The crumbs of Mould sank beneath the water – and as Kun watched, the waters began to shrivel, as if they were being soaked up by a giant sponge. Soon what had been a lake became a waterlogged valley; gradually the marshy bottom dried out as well.

Kun was delighted. He walked down into the valley to inspect the results. Everywhere, the ground was firm and dry. From caves and tree tops, people who had survived the flood began to appear. They looked down into the dry, brown valley, hardly able to believe their eyes. They scrambled down the mountain sides and knelt on the earth, running their fingers through the soil, laughing and crying in relief and delight. Some who had had the foresight to save some seeds began to plant them at once, and others set to work rebuilding their huts on the now firm and solid valley floor. Life had begun again.

Kun smiled, and went on to another valley. Tirelessly, he travelled throughout the land, using the Magic Mould to soak up the floodwaters and create new, fertile fields and valleys.

HEROES AND PROPHETS

There are many legendary figures in the traditions of mankind, heroes who may well have lived but whose deeds have been exaggerated by centuries of storytelling. These men are remembered for their courage in standing up to and defeating the forces of evil, whether by their own strength and cunning or with the help of their gods.

Many of the great religions of the world have stemmed from the lives and teaching of great men, prophets who claimed a unique vision of their God and fearlessly passed on his message to king and commoner alike.

The settlement of the Israelite tribes in the land of Palestine described in the book of Judges, in the Bible, probably took place at the beginning of the Iron Age, around 1200 BC. Palestine was named after a tribe called Philistines; for centuries, the immigrant Israelites battled with their neighbours for survival. This account of David's epic encounter with the Philistine giant, Goliath, goes back to 1000 years BC.

David and Goliath

The land of Israel was being invaded by a hostile tribe called the Philistines, and Saul, chief of the Israelites, sent out a call to all his tribes to send their fighting men to him to meet the enemy in combat. Among the Israelite soldiers were the three eldest sons of a man from Bethlehem called Jesse; he had eight sons altogether, and the youngest son, David, who would normally be at home in Bethlehem looking after his father's sheep, would sometimes be sent by Jesse to take food and supplies to his three brothers with Saul's army.

The Philistine forces were drawn up ready for battle at a place called Socoh in the region known as Judah. Saul and his warriors pitched camp in a valley, and drew up their lines on a hilltop facing the Philistines. Neither side made a move, until there was a stirring in the Philistine lines and a giant of a man stepped forward. He was Goliath, the Philistine champion. He wore bronze armour, and carried a bronze dagger and an iron-tipped spear. A shield-bearer walked before him.

Goliath strode out into the valley between the Philistine and Israelite lines until he was within shouting distance of the Israelites. Then he raised his fist, and bellowed out his challenge to Saul and his men.

'Slaves! Cowards! If you want to settle this dispute, send out your best man to fight against me. If he can beat me in a fair fight, we will withdraw our forces and leave you in peace. But if I should kill him, you will become our slaves. I challenge you! Find a man who will face me, and let us fight it out!'

The Israelites were struck dumb by the challenge. Saul hurriedly called a council of war with the tribal leaders. He wanted to go out and fight Goliath himself. But his advisers would not allow it.

'What would happen to our tribes if you, our chief, should be killed?' they said. 'We would be scattered to the winds, and the Philistines would trample over our land. No, we cannot allow you to go. Brave warrior though you are, you would stand no chance alone against that giant of a man. Who could possibly stand up to him?'

And so the matter stood. The Israelites could find no-one to go out and fight against Goliath, and the Philistines were not prepared to risk an all-out battle against the Israelite army. It was deadlock. Every day, Goliath would stride out and shout out his challenge, strutting up and down between the lines and hurling insults at the cowering Israelites. Every day, Saul grew more desperate, and his authority over the tribes of Israel began to slip away.

One day, David went to the Israelite camp to take some food to his brothers. He arrived just as the army was going out to take up their positions facing the Philistines, and he ran to find his brothers. While they were talking, Goliath, the Philistine champion, came out from the enemy lines as usual and shouted out his challenge. This was the first time that David had heard Goliath, and he was astonished at the reaction of the Israelite soldiers.

'What's the matter with you all?' he asked. 'Who is this foreigner who dares defy the army of Israel? And why does no-one meet his challenge?'

'Look at the size of the man!' retorted an Israelite soldier. 'Saul has offered a rich reward and his daughter's hand in marriage to anyone who will fight him, but no-one will risk his life against a giant like that.'

David's eldest brother Eliab heard David and tried to keep him quiet.

'You don't know anything about military affairs, David,' he said. 'Get back to looking after the sheep, and leave this to us.'

By now Saul had heard what David had said, and sent for the lad.

'What's all this about, my boy?' he asked.

'If no-one else will go and fight Goliath,' David said, 'if no-one else will defend the honour of our god Yahweh and the Israelite people, I'll go myself!'

'You!' exclaimed Saul. 'But you're only a boy. What chance would you have against a giant like that, a trained, professional fighting man?'

'Sir, I'm a shepherd,' replied David. 'I protect my father's flocks against lions and bears. I have snatched a sheep from the jaws of a lion. This Philistine has defied our god, Yahweh. Surely Yahweh will protect me against Goliath, as I believe he protects me against the wild animals of the wilderness.'

Saul was ashamed when he heard David's words, and realized he should have had more faith in the Lord Yahweh himself.

'All right – I agree,' he said. 'Go and meet Goliath – and the Lord be with you.'

Saul sent for his armourer and had David fitted out in a suit of armour and a bronze helmet. Then he buckled his own sword around the boy, and stood back to look at him.

Poor David could hardly move under the weight of the armour.

'I'm sorry, sir,' he said. 'This is no good. I can't fight anyone like this. You'll have to let me handle this my own way.' And he took off the armour, picked up his stick and his shepherd's sling, and went and found five smooth stones from the stream that ran through the camp. Then he set off through the lines in search of Goliath.

Goliath was still pacing up and down, shouting insults at the Israelites. When he saw David coming towards him, he was taken aback, then threw back his head and roared with laugher.

'Look at this!' he shouted to his followers on the Philistine side. 'They think I'm a dog, and they've sent a boy with a stick to round me up!'

Then he turned back to David. His face was dark with rage.

'By the gods of my people, you'll suffer for your impudence,' he roared. 'I'll tear you limb from limb!'

'You stand against me with armour, spear and dagger,' replied David, 'but I come before you in the name of the Lord Yahweh, the god of Israel. Yahweh will destroy you today, and the whole world will know that the god of Israel is a god to be reckoned with.'

With a bellow of rage, the Philistine began to lurch towards David in his heavy armour. David ran lightly towards him, fitting a stone into his sling as he ran. Then he stopped and hurled the stone full into Goliath's face. It struck him on the temple, and he fell with a great crash of armour in front of David.

David stood over Goliath and drew out the giant's sword from its sword belt. He killed the unconscious Philistine with one blow, then cut off his head and held it up for both sides to see. A roar of triumph rose up from the Israelite ranks. The Philistines were terror-struck. They had pinned all their faith on their gigantic champion, and now he was dead their courage vanished. As one man, they turned and ran, hotly pursued by the Israelites.

So David's faith in Lord Yahweh was borne out, and the threat of invasion was over.

The story of Beowulf is set in Denmark, and was probably brought to this country by the Angles and Saxons. The English version is one of the oldest poems in the English language.

Beowulf

THE BUILDING OF HEOROT

In the early days of the Viking kingdoms, Hrothgar the Dane ruled in Denmark. Strong as a bull, yet wise and patient, he ruled unchallenged in a violent world where fierce neighbouring tribes and upstart leaders might burst out at any time, killing, burning and looting.

There were few permanent settlements in the northlands at that time. Men struggled for survival and had no time to plan townships; and in any case they lived in a land of mist and fen, marsh and bog, where the wind howled and hustled and water all too often crept in amongst them across the flat land. They were too busy fighting or building warships to learn about drainage and cultivation; too busy fighting off Geats, Saxons, and half a dozen other tribes large and small who grew strong and flourished for a while, battling their bloody way into their neighbours' land and then declined and withdrew, sometimes to stay quiet for a few years until the next eruption, sometimes wiped out by lurking enemies with a sharp eye for vengeance.

Hrothgar and his Danes extended their power and began to think of building a huge feasting hall, stronger and more permanent than men had seen before. The king and his chief commander had their base camp on a stretch of solid land, free from flooding and good for defence. Beyond it, stretching a long way, was the treacherous fenland where oily marsh gases rose from the spongy land and a man who lost his way would be sucked down into the oozing, porridge-like ground and drowned in mud; that is if he were not first sniffed out by one of the wolves that howled to each other across the flat land, their noise like the wailing of the wind or the cries of pain of the demons and spirits which also haunted the place. For there were loud and persistent whispers of something worse than wolves out there, some shapeless monster that had existed in this bleak deathland since before man appeared; existed in blinding misery and anger which it could slake only for brief moments by acting out the cruelty and death deep within itself. Above all it could not bear happiness and laughter. They were said to make its own pain and misery even more acute, like strong light on a diseased eye.

But King Hrothgar was hardly aware of such fearful gossip. Captive tribesmen were set to build the great hall on dry land. Stonework began to thrust from the ground like strong sprouting new teeth. Masons and smiths and labourers sweated and built for months . . . years until it stood ready: Heorot, the towering hall, a landmark for miles around, like a stone ship, roofed and gabled, bigger than any barn and many times as strong, with a massive wooden door that could resist any surprise attack by enemies; at least, any known enemies. That door, cut from oak and bound with iron, stronger than rocks against water, was vulnerable only to the evil monster, snarling and brooding in its misery deep in the fen, angrily aware of this new thing built on the power and triumph of Hrothgar and his Danes. It writhed and groaned and waited.

Inside the hall was light and laughter, feasting and song. The torches fixed to the walls sputtered and flamed, throwing moving shadows on the stonework. On the rush-strewn floor stood long benches and tables where the warriors feasted, shouting and belching and telling fine stories of their fights. They waved their brawny arms as they boasted and looked back on their violent lives. Yellow manes of hair shook with laughter; rough fingers, calloused by years of gripping sword and axe, splitting skulls or heaving on the oars as they forced their vessels through the bitter cold of the North Sea, now relaxed, curled around their sturdy mead cups. The strong brew and the soft feeling of safety loosened their tongues, and they talked not only of battles and leaders, loot and women, but of the old gods and the new, the worship of Odin and Thor, and the Christian teaching. For this was a time of change: a man might come to accept the new faith in peaceful interludes, but in battle or as tragedy pressed he might still call upon more ancient names.

Hrothgar and his queen Wealthrow sat at the top of the room. The king smiled with satisfaction to see the building finished at last and his trusted followers feasting and lounging at the tables. The huge tapestries stirred on the walls as the wind howled outside, unheeded by the company.

GRENDEL

Across the fens there was a stirring. The monster Grendel dragged himself from the black depths of the lake where he hid in his suffering. He sensed the joy and triumph in Heorot until at last it became quite unbearable. Snarling and moaning he travelled rapidly across the marshes. Limbed like a man, he had a grotesque head and giant strength; his brain was full of cunning and anger.

Hrothgar and his queen had retired for the night, leaving the others asleep round the edges of the hall. There was silence apart from occasional sleep-talk and murmuring and the torches had burnt low. Even the guard nodded.

Grendel hardly hesitated as he reached the huge oak and iron door. It opened before him as if by magic. Quickly he seized the two nearest sleeping thanes, wrenched off their limbs and drank their blood. Cramming lumps of flesh into his mouth, slobbering and moaning with anger, he turned and made off.

In the morning there was despair and misery in the hall. They were used to violence and sudden death, but only from human enemies they could see and recognize and fight against. This latest foe brought disaster like the rest, but came silently in the night and left only huge footprints, odd bits of hair and slime and a smell of death. He was neither human nor animal and they were at a loss. Worse was to follow: Grendel grew bolder and more savage and came again many

times, glutted with the blood of men until it was no longer safe for Hrothgar and his men to sleep in Heorot. They sacrificed to their heathen gods but it made no difference: the killing went on and on and the story of Grendel and his vengeance spread across the Northern world.

It came to the ears of Beowulf, a distant relative of Hrothgar, who ruled the Geats. He was already famous for his strength and heroic deeds and he decided to go to Hrothgar's aid. With thirty followers he built a ship, launched it beneath the cliffs and set off. Soon they reached the Danish coast and were given safe passage to Heorot by the watchmen guarding the shores, leaving their ship riding at anchor. Their corselets chinked and gleamed as they travelled the stony road till they arrived at Heorot. There they propped their shields and ash-spears against the walls of the building and rested, as Hrothgar questioned and welcomed their leader, whom he had seen only once before when he was a child.

Beowulf spoke of his victories; he told how he had destroyed sea monsters and giants, and asked to be allowed to take on Grendel without weapons, relying on the strength of his arm alone. He and his men would sleep in the hall that night, as no one had done for several years since the terror began.

There was some jealous opposition to Beowulf's plan. But the king was glad enough to agree to it, and later that evening Beowulf and his men settled down for the night round the walls of Heorot, waiting and listening.

Far off in the fen, Grendel knew that someone had dared to return at night to Heorot. He smelt or felt the new presence and fury welled up in his brain like a foul mist. In the dark night, as wolves howled and the creatures of day hid themselves, Grendel came gliding across the land. God's curse sat heavily on him and he ached to kill.

Once again the door sprang open at his touch and he stepped inside. An awful light shone from his eyes like a flame and he was aware of the company of men – all asleep except one. It made him ravenous and at once he seized and ate the nearest man to him. Then it happened: the next man was Beowulf. As Grendel groped for him he found his hand seized and locked by a grip such as he had never known. Terror flooded into him and he tried to break away. But it was no use. Beowulf sprang up and hung on with all his strength. The two of them swayed and staggered across the hall, well matched for strength. Benches, tables and mead cups crashed to the floor, tapestries were ripped down. Alarmed by the hideous howling of the monster and the crashing of furniture, Beowulf's men sprang up to help him, seizing swords and spears wherever they could. But weapons were useless. Grendel could not be touched by them; they broke like matchsticks. It was a battle of strength and will-power. With a last desperate effort, Grendel tore himself away. As he did so, his arm was wrenched from its socket at the shoulder, the joints burst and with the shriek of a soul in hell he bolted across the moors mortally wounded, dripping blood and in terror of death. He dragged himself back to his retreat deep in the mere, where the water surged with his blood as he returned to die beneath the marsh.

Beowulf meanwhile, exhausted but triumphant, hung up the huge bloody arm and hand as a trophy from the roof beam, and people came from miles around next day to see it. There was great rejoicing in Heorot. Warriors came to praise Beowulf and to gaze on the huge limb, hanging from the rafters, the blood clotted and dried at the shoulder joint and the fingers curled in agony. Hrothgar organized a huge feast and precious gifts were showered upon Beowulf and his men as light and laughter returned after nightfall for the first time for many years.

THE FINAL BATTLE

But Grendel had not lived alone. Tales had been told over the years of two hideous monsters in the marshes. One was like a deformed woman, and both were thought to live in a region of wolf-haunted slopes, windy headlands and fen, where the hill stream fed into a lake of immense depth, at the foot of a cliff. The water was overhung by ancient twisted trees, the air was dank and gloomy and the oily waves kept the surface always restless, always in motion. From her dwelling in the depths, Grendel's mother came panting to avenge her son, seized a sleeping Dane and made off before she could be touched. She carried off her son's arm as she fled, whining in angry grief.

Hrothgar quickly summoned Beowulf and told him the bad news. One of his closest and most trusted kinsmen had been carried off, snuffed out like a candle as he slept. Beowulf knew that he must follow Grendel's mother to her lair and put an end to the terror once and for all. He and his troop armed themselves and set out.

It was not difficult to follow the monster's track. Before long they came to the lake. It was cheerless and menacing, but worst of all was the sight of the latest victim's head twisted off and discarded at the water's edge. The water was bloodstained.

Now Beowulf got ready for the final battle. He put on his closely woven armour to protect him from weapons and the attack of the monster. His helmet and Hrothgar's famous sword of victory completed his equipment. Hrothgar had ridden with him to the lake's edge and Beowulf now asked him to protect his men and send his treasure back home if he should not return. Then he plunged into the lake.

Because of his supernatural protection in his fight against Grendel and his mother, Beowulf was able to travel beneath the water without drowning. It took him several hours to reach bottom and long before that Grendel's mother was aware of the invader swimming down like a marauding shark seeking her out. No sooner was he there than she seized him in her claw hands, eager to tear him in pieces. But now the ring chainmail of his armour protected him. Her fingers groped and clawed for a chink but it was no use. So she carried Beowulf off to her lair in the heart of the mere.

He saw dimly that he was in a kind of underwater cave, a hall where the water did not come. A cold, pale light showed him Grendel's mother as she flung him down. At last he could use his sword and brought the blade down with all his strength onto her head, heavily as a crashing pine tree. But it failed him. For the first time that famous sword was turned aside as if it had been matchwood, such was the evil power of Grendel's mother.

Beowulf was not put off. He realized that he must trust in

his own strength and he seized her and flung her to the floor. They struggled there like two giants, writhing, gasping and searching for a death grip. Grendel's mother seized her opportunity and drew her dagger, broad and bright edged, flashing like a jewel in the half light. Eagerly she lunged at him, longing for vengeance at last. But the blade was turned aside by his corselet.

Then Beowulf saw on the wall of the cave an ancient sword, forged by giants many ages ago, huge and deadly, with a chain hilt and patterned blade. Seizing it, he struck despairingly at his enemy's neck. Her spine broke and the sword ran through her dying body as she fell to the floor in her agony. As she died a cleansing light spread through the cave like the sun emerging from clouds.

Gazing round, Beowulf could now see the many ancient treasures hoarded in the cave on the walls and the floor. And there in a corner lay Grendel's corpse, with a gaping wound at his shoulder. Triumphantly Beowulf cut off his head for a trophy.

At the lake-side the watchers were alarmed by the surging blood in the water. Many hours had passed since Beowulf left and they were afraid that he had died, savaged and destroyed under the windy water by the lurking fiend in her own home. Hrothgar and his men departed sorrowfully. Only Beowulf's own men stayed on, hope wilting like a dying plant in their hearts.

Deep in the lake, the sword that had killed Grendel's mother melted before Beowulf's astonished eyes, as her blood poured out, acting like acid on the ancient metal. Only the hilt was left and Beowulf seized this and the severed head before striking out for the surface. As he broke surface in the now clear water, his men cheered with relief and dragged him exhausted to the shore. They stripped him of his armour and rode back in triumph to Heorot, four of them carrying Grendel's head on their spear-shafts. When they arrived they carried it by its lank, weed-like hair to the king's dais, whilst his men gazed in horror and loathing at the evil features.

The terror was over. After the feasting and laughter, Beowulf and his men sailed back home, laden with gifts and treasure.

The story of St. George is usually thought of as an English legend, but the real St. George probably lived in Asia Minor, the country we know today as Turkey, in the third or fourth century A.D. He was a Christian who was later martyred for his faith. This story was probably brought back to England by the returning Crusaders in the twelfth century A.D., and shows St. George depending upon his faith in God, as did David when he faced Goliath.

St. George and the Dragon

The city stood beside a lake; the lake was huge, fathomless, deep as the sea. In its dark depths a dragon lurked. Every day it heaved itself from the mud and slime of the lake-bottom and surged into the city, dripping rancid water and strands of weed. It prowled the streets for prey: dogs, cats, people were snapped up and gulped alive.

There was no defence. The king's army galloped out to do battle, and slunk whimpering back; the people prayed to their heathen gods, and their gods were deaf. The dragon grew fat; the city shrank.

At last the king proposed a dreadful remedy. 'My people,' he said, 'if we feed the monster on our children, perhaps it will be satisfied. Send out your children, one by one; and on the last day I'll send my only daughter, jewel of my life.'

So the people did. Each day they watched as another group of children set out on their last, hideous journey. No games; no singing; no laughter. The schools emptied; the streets were silent; the houses were quiet and still.

At last there were no children left. It was the turn of the king's own daughter. Weeping, he dressed her in wedding clothes, for her marriage to death. The people went to the lakeside in silent procession, to see her die.

Now it chanced at this time that St. George was travelling in these parts. He came down to the lakeside so that his horse could drink, and found the silent crowd and the girl in wedding clothes, sitting forlornly on a rock.

'My child,' said the saint, 'what is it? What's happening?'
Weeping, the princess explained.

'Why don't you pray to God for help?'

'Our gods watch us die, and laugh.'

Then St. George lifted his eyes to heaven and prayed aloud to the Christian God. 'O Lord God all-powerful, send a sign. Grant me strength to crush this dragon and win the people's hearts.'

There was a stillness, silence over the water; the world held its breath, and out of the stillness came a voice. 'Fight, George. My strength is yours. Fear nothing.'

Then, as the people watched, the lake water began to heave and churn. The dragon was stirring. 'Go, my lord!' cried the girl. 'Save yourself while still there's time!'

But St. George held his ground. As the dragon's head, hung with green weed, broke surface and surged towards the shore, while the people flinched and ran, he made the sign of the cross in the air and flung a stone hard in the monster's face.

The dragon gave a hiss and bent its head. At once St. George put his foot across its neck where it arched and snaked in shallow water. 'Quick, child!' he shouted to the girl. 'Take off your belt and give it me!'

The girl did as she was told. Praying to the Holy Ghost, St. George lashed shut the dragon's jaws and made a lead for it as if it was a puppy dog. 'Take it,' he said to the girl. 'Lead it to the market square; I shall kill it there, in the name of God.'

The people fell back as the tiny girl led the dragon after her, up the beach and into the city. It was dazed, obedient as a lamb. They came to the market square, and with a single silver blow St. George cut off its head. The people surged round, kissing his feet for joy. And before that day was done, they were all converted by the miracle, and worshipped the Christian God.

King Arthur and the Knights of the Round Table are the heroes of one of the greatest legends of Britain. We do not know whether there ever was a real King Arthur. If he did exist, it was probably in the fifth century A.D., and Arthur may have been a leader of the Britons in their wars against the invading Saxons.

Some of the most famous stories of King Arthur's knights are about the Quest for the Holy Grail. Versions of these stories are found throughout Europe, and the author who did most to sort out these various legends and retell them in English was Sir Thomas Malory, in the fifteenth century.

The Quest for the Holy Grail

When Jesus was crucified one of those who mourned him was Joseph of Arimathea. In his grief he longed for something he could keep which had belonged to his master. On the day of the crucifixion he brought a vessel of rich beaten gold and caught in it drops of Christ's blood as it dropped from the wounds. Afterwards, Joseph took the Grail from Calvary and hid it in his own home, where he prayed before it every day. In time, the Jews who had been hostile to Jesus became suspicious of Joseph and spied on him. They found out what he was doing and he was arrested and imprisoned.

The prison was a high tower, its walls tougher than any oak tree, its windows small and the door closely guarded. But none of this was enough to keep Joseph inside. He prayed to God to release him and preserve the Holy Grail; at once the walls crumbled like sand and he returned to his home.

His enemies were furious. They were determined to be rid of him and their leaders banished him from the land, together with Nicodemus, another follower of Jesus. Together, Joseph and Nicodemus took ship with some friends and entrusted themselves to God, taking with them the Grail. After many days they came to the island of Britain where they built themselves a refuge and decided to stay. However, they were not left for long in peace. The local tribesmen attacked them, and they were close to starvation and defeat when Joseph called on God to help him by means of the Grail. All the company sat down in due order at empty tables at Joseph's instruction. Then he blew a horn, and at the sound, the Grail appeared miraculously, pouring out wine for everyone there and causing rich, steaming dishes of the choicest food to appear as if by magic at the empty tables. The company could only gaze in astonishment as this happened; but soon their astonishment turned to worship of God's power and mercy.

By this and other miraculous means, Joseph and his friends were protected until his death. The Grail itself was not kept in the possession of any one man, but it was revered as a sacred object and it was thought that only the most saintly amongst Joseph's descendants might ever see it.

★   ★   ★

More than four hundred years passed. King Arthur and his knights held court at Camelot. In order to avoid jealousy amongst his knights, Arthur had had a huge circular table made so that everyone should be equal and no one feel inferior. Around it stood heavy oak chairs with high backs for the knights to sit in. It was the feast of Pentecost at Whitsun and it was midday. On their return from Church, relaxed and talking to each other quietly, the knights entered the great hall where the Round Table stood. Their appetites were sharp and they looked forward with pleasure to the coming feast. As they approached the table they were amazed to find that each chair had words in letters of gold written across the back, forming the name of one of the knights who was to occupy that place. Their talk turned to exclamations of surprise as each man moved round the table, peering at the rich golden words and marvelling at what they saw.

One of the seats was different from the rest. On it was written: 'This place will be filled four hundred and fifty years after the Crucifixion of our Lord.' Sir Launcelot mused for a while when he saw this, then exclaimed: 'I believe this is the very day, for it is exactly four hundred and fifty years since our Lord was crucified. If you all agree, let us cover this seat until the arrival of whoever is to sit in it.'

This was agreed by all and the chair was draped with a rich silken cloth.

King Arthur would have begun the feast but someone reminded him that it was customary for him to see some strange happening or object before they ate. 'You are right,' replied the king. 'The safe arrival of Sir Launcelot and his cousin at court quite put it out of my mind.'

At that moment, a squire of the court came into the Hall. 'My King,' he called, 'I bring you news of something very strange.'

'What is it?' asked Arthur.

'Below at the river I have seen a great block of stone floating on the surface. What is even more strange is the sword which is plunged deep into it.'

'I will go and see for myself,' said Arthur.

All the knights left the hall and hurried to the river bank. Sure enough there it was. A block of what looked like red marble floated on the surface, rocking gently with the motion of the water. They could see the little veins in the marble, running criss-cross through the stone. In the middle of it a sword was embedded for almost half its length. And what a sword it was! The burnished silver blade reflected the bright sunlight which winked and gleamed with the rocking motion of the water. But richest of all was the belt; red, green and gold precious stones shone from it, and on it was written in golden letters: 'I shall never be drawn out except by the man who will wear me, and he shall be the greatest knight in the world.'

King Arthur turned to Launcelot. 'This sword is meant for you, for I am sure you are the greatest knight in the whole world.'

'Not so,' answered Launcelot sadly. 'It is not meant for me and I have not the courage to try to draw it. Let me tell you also that anyone who tries to draw it out and fails will one day be wounded mortally by that same sword. Sir, I believe this is the day on which the search for the Holy Grail shall begin.'

Arthur turned to Sir Gawaine. 'Good nephew, make the attempt, for love of me.'

'Forgive me, I will not,' answered Gawaine.

Arthur was angered by this reply. 'I command you to draw the sword.'

'I obey your order,' answered Gawaine. But he could not move the sword for all his strength.

'Sir Gawaine,' said Launcelot. 'You may be sure that this sword will one day wound you so deeply that you will wish you had never laid hands on it, not for the greatest castle in the kingdom.'

'Sir Launcelot,' answered Gawaine, 'I could not disobey my uncle's orders.' When Arthur heard this he was sorry and he asked Sir Percival to try his hand at it, for love of him.

Sir Percival replied, 'Willingly, to keep Sir Gawaine company!' He grasped the hilt and pulled with all his strength, till the sweat stood on his forehead and his muscles quivered with the effort; but it did not move at all. It was as if it were part of the rock itself.

Several other knights followed suit but had no more success. At last Sir Kay said to Arthur, 'We have seen a marvel. Now we can sit down and eat.'

So they returned to the hall. Each man knew where to sit and soon all the seats were filled except the one previously set aside. The young squires of the court hurried in and out waiting on the knights and all was bustle and noise. The clattering of platters and the hum of voices rose to the beams in the roof of the great hall.

All at once there appeared at the door an old man, dressed in white. With him was a young knight. He had no shield or sword, only an empty scabbard hanging by his side. The old man spoke. 'Here is a young knight of noble birth, descended from Joseph of Arimathea, come to fulfil the destiny of this court.'

'You are most welcome,' said Arthur to them both.

Then the old man asked the young knight to remove his armour. The knights saw that he was dressed in a red silk garment, over which was a cloak lined with white ermine.

'Follow me,' said the old man, and led his companion to the empty place which was next to where Sir Launcelot sat. As he drew aside the silk cloth there was a gasp from the other knights in the hall; on the chair, in golden letters, was written, 'Here is Galahad's seat.'

'Sir,' said the old man, 'this place is yours by right.' The young man took his place with assurance.

'Now you may leave, for you have done everything you were commanded,' said Galahad to the old man. 'Greet King Pelles, my grandfather, for me, and also the Fisher King, and tell them I shall return as soon as I can.'

The old man turned away. At the door, twenty noblemen met him, mounted on fine horses that tossed their heads and whisked their tails in the sunlight. When he had mounted his own horse, they trotted away, their iron horse-shoes clattering on the hard stones of the courtyard.

King Arthur's knights gazed in awe at young Sir Galahad; they marvelled at his youth and courage, that he dared to sit in that seat. They could only think he was sent by God. Whispering to each other they said, 'This must be the man who is to succeed in the search for the Holy Grail. No-one else ever sat there through all the ages.'

Galahad sat calm and unmoved. Sir Launcelot looked at him and his heart swelled with pride; for this was his own son.

King Arthur welcomed the young knight. 'You will inspire many knights in their search for the Holy Grail, and shall yourself succeed where all others have failed.' Then he took Galahad to see the marvel of the sword set in the stone. When Queen Guenevere heard what was happening she also went down to the river with her ladies and showed them the great block, rocking quietly on the water.

'This is as wonderful a sight as I ever saw,' said Arthur. 'Many strong knights have tried to draw out the sword and failed.'

'That is not surprising,' replied Galahad. 'It is my task, not theirs. Because I knew about this sword I brought with me only an empty scabbard, as you see.' He grasped the jewelled sword-hilt firmly and with an easy movement, drew it out as if it were embedded not in marble but in sand. The blade flashed briefly in the sunlight until Galahad sheathed it in the scabbard. 'Now it is properly filled,' he said.

Then King Arthur, knowing that his company would soon leave him in the quest for the Holy Grail, asked them all to assemble in the field for jousting so that he could look back on the day after they had all gone. This they all did and fought in the lists before the king and queen and all the court.

Trumpets blared, horses neighed and the knights clashed together in combat. Some were unhorsed, tumbling noisily from their sweating steeds, and many a lance shivered in pieces, but Galahad, though he refused to carry a shield, proved his knighthood against all comers with his new sword. Only Sir Launcelot and Sir Percival could stand against him.

So they went to the great church and afterwards in to dinner, each man sitting in the place he had occupied before. Suddenly the air was split with the crackle and boom of thunder; so much so that everyone was afraid the building would collapse. At the height of the tumult, as the thunder boomed like cannon, there burst upon them a piercing light, a sunbeam seven times brighter than any man had ever seen. Each man there was touched by the spirit of the Holy Ghost. They looked at each other in wonder and saw their old companions more perfect in appearance than they had ever been. For a while they were struck dumb by all this and could only gaze silently around at the rest of the company in awe.

In the electrified silence they were aware suddenly that the Grail was in their midst. It moved into the hall, apparently of its own accord and covered over with a white silk cloth. As it moved, the whole hall was filled with delicious smells; the whiff of succulent meat, spiced wines and fresh baked bread all teased their nostrils and drew them on to eat. Every one of them tasted such food and drink as he had never known. When they had had all they could wish for the Grail disappeared, none knew how. At once their tongues loosened and they could speak again.

'Praise be to God,' said the king, 'for what he has shown us on this Holy Feast Day.'

Gawaine replied, 'We have been refreshed with whatever food and drink we wished for. The one thing we were denied was any sight of the Grail itself beneath its covering. Therefore I take a vow that tomorrow morning without delay I shall set out in search of it for a year or more if need be. I shall not return to court until I have seen it clearly: if I do not succeed, I will return only if it is God's will.'

Hearing this, most of the rest of the company stood up and vowed to do the same.

Arthur was displeased at this turn of events. Tears rose in his eyes and he reproached Gawaine. 'You have filled me with sadness, for I fear that the knights of the Round Table will never meet again after this.'

'Comfort yourself,' said Launcelot. 'We set out on a most noble quest. We know death will come to all; and in what better way could we meet it?'

'Launcelot,' said the king, 'it is the great love that I have for you all that makes me speak out so. No Christian king ever had such a noble company at his court, and that is why I grieve.'

When the ladies heard they also were saddened, especially Queen Guenevere. 'I am surprised that my lord will let them go,' she said. Many of the ladies would have gone with their own knights, but it was forbidden them.

The next morning, the knights put on their armour, heard Mass and solemnly swore to follow the quest. And so they rode out on many paths, into the wide world, seeking the vision of the Grail. Through forests, swamps and over many a wide sea they made their way. But only Galahad achieved it, and that on the point of death. Arthur's fears were proved true. The fellowship of the Round Table was at an end.

We have already seen one account of the god Vishnu taking human form as a little boy, Krishna (page 60). Another great Hindu epic story is the Ramayana, which relates how Vishnu took the form of a prince named Rama.

This part of the story tells how Rama defeated the demons Maricha and Subahu, and how he won his bride, Sita.

Rama and Sita

On the banks of the great river Sarayu at the foot of the Himalayan mountains there was once a fabulous kingdom called Kosala. Kosala was famous throughout the world for its beautiful capital city which had been built in the earliest times by Manu, father of mankind, after his miraculous escape from the flood. The land was ruled by King Dasharatha, who was a wise and thoughtful king, and under his leadership Kosala became a rich and happy country. There was only one thing Dasharatha lacked: a son to inherit his kingdom.

One day Dasharatha sent for the Brahmans, his household priests, and gave them orders to prepare a sacrifice to the gods in the hope of persuading them to send him a son. The Brahmans found a good place for their ceremonies, and soon the air was filled with the sweet smell of the sacrificial offerings and the sound of the priests' chanting. As the king and his followers watched, Agni the god of fire appeared, and a whole host of lesser gods. They spoke solemnly with one voice, as the king and his court listened in awe.

'Ever since the demon Ravana appeared on earth, the life of the gods has been troubled by his mischief, for none of us can control him. May the great Lord Brahma, creator of the universe, grant that a son be born to King Dasharatha who will be powerful enough to defeat the demon Ravana and give us back the peace we once enjoyed.'

As the royal court watched, the ranks of the gods parted

and a beautiful spirit creature stepped forward, carrying an exquisite goblet full of a mysterious liquid. The spirit spoke directly to the king:

'Take this goblet, King Dasharatha of Kosala, city of Manu, and give it to your wives that they may drink. Then you will father the sons that you desire, and one of them shall rid us of the curse of Ravana.'

The king did as he was commanded, and sure enough a year later his wives gave birth to sons: Rama, Bharata, Lakshmana and Satrughna. The four half-brothers grew up to be close friends, but Rama's closest friend and companion was always Lakshmana.

The time came when Rama was old enough to get married. King Dasharatha was consulting his chief priest about the matter, when there was a visitor to the court. It was Vishvamitra, a famous warrior who had given up the arts of war and become a Brahman priest. He was given a welcome fitting to one of his caste, and in due course he explained to King Dasharatha the reason for his visit.

'My religious devotions, oh King, are as you know very important to me, and in order to concentrate upon my prayers, I must have absolute peace and quiet. Just lately, however, my calm has been disturbed by two evil demons, Maricha and Subahu. Now that I have become a Brahman priest, I may not attack them and destroy them as I once would have done.'

Here the aged priest paused, and fixed his burning gaze upon the king. Dasharatha shifted uneasily in his royal chair, and waited for Vishvamitra to continue.

'You can help me, oh King, by sending Rama, your son, to destroy these two evil demons as I would have done in the days of my youth.'

The king was taken aback at this request.

'I do not see how Rama, who is still but a youth, can help you in such a quest, most worthy Vishvamitra. He has not even completed his military training.'

Vishvamitra turned his serene gaze upon the king's chief priest, who hurried forward and whispered into the king's ear.

'Remember, oh King, that Vishvamitra was once a great

warrior. He will, I am sure, watch over your son and instruct him in the necessary arts to destroy these demons. And who knows? Vishvamitra has much influence in the kingdoms of India. Perhaps he will interest himself in the problem we were discussing, and in his gratitude help us to find a suitable wife for your royal son.'

King Dasharatha saw the wisdom of his chief priest's words, and sent for Rama and his companion Lakshmana at once. After the offering of sacrifices to ward off evil spirits, the two brothers set off with Vishvamitra in search of the demons. They slept that night on a bed of leaves on the banks of the river Sarayu, and in the morning, Vishvamitra began their training.

When Rama and Lakshmana had acquired the skills of fighting and the strength of warriors, Vishvamitra was satisfied. He called Rama to him, and gave him a magic potion which would give him superhuman strength. Then he pointed to the forest near their camp.

'Over there, in that forest, dwells an evil demoness. She is called Tataka, and she is the mother of the two demons who torment me. She is evil and merciless, and before we seek out her demon sons, you must destroy her.'

Rama saw the force of Vishvamitra's words, and taking his bow, went with Lakshmana to the very edge of the forest. There he held his bow aloft, and twanged the bowstring as a challenge to the demoness. When Tataka heard the sound and realized that someone had dared to challenge her, she came rushing out of the forest screaming with fury. Rama stood his ground, fitted an arrow to his bow, calmly took aim, and shot the demoness through the heart.

'Well done, noble Rama!' cried Vishvamitra, who had followed the brothers to see how they came through this first test of courage. 'Now we shall seek out Maricha and Subahu. Come, follow me.'

The three set off through the forest, ever watchful lest the evil demons should catch them unawares. On the sixth day, they were suddenly confronted by the two demons, who rushed at them with bloodcurdling cries and whirling swords to avenge the death of their mother, Tataka.

Rama was ready for them, and quickly hurled a sharp

javelin at the first demon, Maricha. It struck him full in the chest, with such force that he was lifted bodily and carried to the sea, and seen no more. The second demon, Subahu, blind with fury, was almost upon them when Rama picked up a club and threw it straight at him. Subahu was killed instantly by the force of the blow, and crashed to the ground at their feet. Their mission was fulfilled.

The aged Vishvamitra came hobbling up to Rama and flung his arms round his neck.

'You have fulfilled your quest, great Prince, and you shall have your reward. Come with me now to Mithila, the city of King Janaka. There we shall find a contest worthy of one as mighty as you.'

Rama and Lakshmana were puzzled at the sage's words, but their father had instructed them to go with Vishvamitra and they followed hin dutifully. They travelled for many days through the forest, and eventually came to the city of Mithila.

When King Janaka heard that the great Brahman had arrived in his city with two young warrior princes, he hurried to meet them.

'Greetings, most noble Vishvamitra!' he cried, bowing deeply before the old priest. 'You come to my humble city at a most opportune moment, for I am holding a contest to find a husband for Sita, my only daughter. You shall help me judge the contest.'

Vishvamitra bowed in return, and indicated that he would be honoured to assist Janaka in the search for a son-in-law. They went at once to the competition arena, and on the way King Janaka enquired about Vishvamitra's two distinguished-looking companions.

Vishvamitra explained that Rama and Lakshmana were the sons of King Dasharatha, and recounted their exploits and the destruction of the demons. By now they had reached the arena, and there, in the centre of the field, was an enormous bow, the height of two men; the great Bow of Mithila.

A hush fell upon the crowd as King Janaka entered the arena with the three adventurers. The king stood before the bow, and addressed the crowd.

'This is the great bow of Mithila, which no man has ever

bent or strung. If the prince Rama is able to bend and string the bow, then I give him my daughter Sita in marriage.'

A great roar went up from the crowd, and Rama stepped forward. The noise subsided, and Rama's companions watched, tense with expectation.

Rama picked up the bow, placed one end upon the ground, reached up, and pulled the other end of the bow with all the superhuman force which Vishvamitra's magic had given him. The great bow creaked and groaned as Rama forced it to bend, then fitted a bowstring to it and began to pull it back. Suddenly, there was a crash like a thunderclap, and the bow was in two pieces. Not only had Rama succeeded in bending and stringing the bow, he had bent it so far that it had broken.

King Janaka stepped forward, shaking his head in wonder.

'Now I have seen a strength and courage I would not have believed possible,' he proclaimed. 'Truly, Rama, son of Dasharatha, descendant of the gods, is a fit husband for the Princess Sita. Let her come forth.'

At these words, the ladies-in-waiting brought the blushing princess into the arena and presented her to Rama. Shyly, she knelt before him, while the crowd roared their approval. Rama lifted her to her feet, and gravely thanked King Janaka for the great honour he had received.

So Rama found a bride, and Sita a husband, and both King Dasharatha and King Janaka were happy, thanks to Vishvamitra the Brahman.

Before the Israelites settled in Palestine, they had been slaves in Egypt. Moses, who was an Israelite, had been brought up as an Egyptian prince, but had killed an Egyptian overseer and had been forced to flee to another country, Midian. Most scholars believe that Moses lived in the thirteenth century before Christ.

This account of the call of Moses, from the third chapter of the book of Exodus, is regarded by many Jews, Christians and Muslims as historical, not legendary.

Moses and the Burning Bush

Moses was working as a shepherd, looking after the flocks of his father-in-law, Jethro. The country of Midian was bleak and barren, and in order to find grazing for the sheep, Moses led the flock for miles along the edge of the rock-desert until he came to Mount Horeb. This was an awesome, craggy mountain with steep cliffs and deep ravines; the Midianites called it the Mountain of God.

While the sheep were quietly grazing on the sparse grass on the mountain slopes, Moses sat down against a rock to rest his feet. It had been a hot, tiring walk, and he was exhausted. Just as he was closing his eyes, something bright caught his attention. Sitting up, Moses looked hard to see what it was.

Just a short distance away from him was a thorn bush, one of many on the mountainside. This one, however, seemed to be on fire. At first, Moses thought it might have caught fire spontaneously in the hot sun; then he noticed something strange. Not only was the fire not spreading to any of the other bushes around, but the bush which was on fire was not being burnt away. It seemed to be surrounded by a very intense, white light, and as Moses watched, he knew this was no ordinary experience.

He began to walk slowly towards the burning bush, like a man in a dream. As he walked, he heard a voice calling his name.

'Moses, Moses.'

The voice seemed to come from the centre of the burning bush. Moses stopped and answered.

'Here I am.'

'Come no nearer, Moses. Take off your sandals, for the place where you are standing is holy ground.'

Moses at once obeyed, trembling with fear. What could this mean?

'I am the God of your forefathers,' the voice continued, 'the God of Abraham and Isaac and Jacob.'

Now Moses understood. He knelt down and covered his face with his hands, afraid to look any more at the burning bush.

'I have seen the misery of my people, the Israelites, and their suffering in Egypt,' God said. 'I am going to rescue them from slavery, and lead them to a new land, a land where milk and honey are plentiful, a land that will be their own. You, Moses, are to return to Egypt and lead the Israelites out of slavery.'

Moses felt his heart sink. How could he possibly go back to Egypt? He was on the run from the Egyptians – he would be arrested at once. He began to protest.

'But Lord, the Pharaoh will never listen to me,' he said. 'He will never let the Israelites go.'

'You will not be alone, Moses. I am with you. You will lead the Israelites out of Egypt, and bring them to worship me here on this mountain.'

Moses could see there was no way out. Desperately, he appealed to God.

'How shall I convince the Israelites?' he asked. 'They'll never believe that you really spoke to me.'

'What are you holding in your hand?' the quiet voice from the bush asked.

Moses looked down. 'A staff,' he replied.

'Throw it on the ground,' God said.

Moses obeyed. At once the staff turned into a deadly snake, writhing and darting its forked tongue. Moses leapt back in alarm.

'Now reach down and pick it up by the tail,' the voice commanded.

Moses was terrified, but something made him obey. He

stepped over to the venomous-looking snake, and took hold of it by the tail, expecting it to turn and bite him. To his relief and astonishment, the snake at once turned back into a staff.

'You see – you will have no difficulty in convincing your fellow-Israelites that it really was their God, the God of Abraham and Isaac and Jacob, who appeared to you.'

One by one, Moses' objections were being swept aside. He had one more despairing try.

'But, Lord,' he said, 'I am no good at public speaking – I never have been. I stammer, and I can't make myself clear.'

'Think, Moses,' the voice replied. 'Who is it that gives man the power of speech? Who gives man sight and hearing? Is it not I, the Lord God? Go – I will tell you what to say.'

'Oh, please Lord,' wailed Moses. 'Send anyone but me!'

At this the voice in the bush grew stern.

'Now listen, Moses. You have a brother, Aaron. He is already on his way to Midian to meet you. Go and find him, and he can be your spokesman to the people. Go now – and don't forget to take your staff, because you'll need that to perform the signs to convince the people.'

Slowly, the light surrounding the bush faded. Moses stood up, rubbing his eyes; he rounded up the sheep and set off as quickly as possible back to Jethro, to tell him that he must leave at once for Egypt.

*Many times in the history of Israel, the Old Testament tells us, the
people turned away from their God, Yahweh, and worshipped other
gods. Elijah, one of the earliest prophets of Israel, who lived in the
ninth Century B.C., here challenges the prophets of the god Baal.*

Elijah and the Prophets of Baal

Lord Yahweh had spoken, and his words were clear:
'Hear, O Israel! I am the Lord your God, and you shall
have no other gods before me.' While the people remembered this commandment, they prospered; but when
they broke it, they were punished with disaster and suffering
till they changed their ways.

The commandment was broken by Ahab King of Israel
and his Queen Jezebel. They forgot Yahweh, and instead
worshipped Baal, a spirit-god from another land. They built
him temples, and all the people bowed down and worshipped him.

Only one man remembered Yahweh: Elijah the prophet.
He went before King Ahab, and said, 'Your people have
sinned, and they must be punished. In the name of Yahweh,
the living god I serve, there will be drought and famine over
all your land. From this day on, there will be neither dew nor
rain.'

Elijah went to live by a brook called Cherith, in a distant
part of the kingdom. Every night and morning, Yahweh's
ravens brought him bread and meat, and he drank from the
brook.

Over the rest of the land, famine ruled: famine, pestilence
and drought.

When the time was right, Elijah went before King Ahab
again. Ahab said, 'Is it you again, the troubler of Israel?'

Elijah said, '*You* are the troubler of Israel, not me: you have

forgotten Yahweh, and worship Baal instead. Now the time
has come to end the matter once and for all. Send messengers,
and call the people of Israel together on the slopes of Mount
Carmel. And call together the prophets of Baal too: four
hundred and fifty of them, and the prophets of all the lesser
gods.'

Ahab did as he was told. His people gathered on the slopes
of Mount Carmel: wretched, parched with drought, starving. The prophets of Baal gathered too: four hundred and
fifty of them, with the prophets of all the lesser gods. They
stood in a dusty, distrustful crowd, waiting to see what Elijah
was going to do.

Elijah stepped forward. 'People of Israel!' he shouted.
'How long must this go on? How long till you make up your
minds? If Yahweh is God, then follow him and no other; if
Baal is God, forget Yahweh and follow him.'

No one answered. The people stood dejected in the hot
sun. They looked down at the dust.

Elijah spoke again. 'Let this be the test,' he said. 'I am one
man, one man only, the only prophet of Yahweh left in
Israel. There are four hundred and fifty prophets of Baal, and
the prophets of all the lesser gods. Bring out two bulls; one
for them, one for me. Let them take their bull, cut it up and
place it on the altar – but light no fire underneath. I will do
the same with mine. Then they can call on Baal to send down
fire; I will call on Yahweh. The god that answers by fire – let
him be your god, and him alone.'

There was a murmur from the crowd, low at first, then
louder. A murmur of agreement.

The bulls were brought. The prophets of Baal chose one,
and the other was put aside for Elijah. The prophets of Baal
built a huge pile of wood for the sacrifice. They killed the
bull, cut it up and placed it ready on the wood.

When everything was done, they began their prayers.
They called Baal's name; they shrieked and cried to him to
send down fire. They danced round the altar, chanting and
beating drums. All morning they leaped and shrieked and
prayed – till it was noon, and the sun was high.

Then Elijah began to mock them. 'Louder!' he said. 'Shout
louder! He's a god, after all. Perhaps he's thinking, or in the

lavatory, or gone for a walk, or asleep. Shout louder and wake him up!'

The prophets cried even louder, danced even more furiously, and began gashing themselves with knives and skewers, as their custom was.

But nothing happened. No answer, no voice, no fire.

Then Elijah said to the people, 'Gather round.' And they gathered round.

First, he repaired a crumbling, broken altar made on the mountainside to Yahweh many years before. He finished it off with twelve stones, one for each of the tribes of Israel. He dug a trench all the way round it, a single metre deep and a metre wide.

When the altar was ready, he piled it high with wood. He killed his bull and laid it on top.

Then he said to them, 'Fill four barrels with water and pour it over the sacrifice, and over the wood.'

Four barrels! In a time of drought! Grumbling, they obeyed.

'Do it again,' he said. 'And again.'

The water ran down, soaking the altar and the wood, and filled the trench. The people watched, silent, to see what would happen.

At last, at the time of the evening sacrifice, Elijah the prophet went and stood by the dripping altar. He began to pray in a quiet voice. 'O Lord Yahweh, God of Abraham and Isaac, make it plain today that you are the God of Israel, that I am your servant and that all this was done according to your word. Hear me, O Lord, hear me: show the people of Israel that you are God, and turn their hearts back to worship you again.'

There was a blaze, a hiss, a crackle of light. The fire of Lord Yahweh fell and consumed the bull, the altar stones, the dust, and licked up the water that was in the trench.

The people fell on their faces in the dust. 'Lord Yahweh is God,' they cried, 'Lord Yahweh is God!'

And Elijah said, 'Take the prophets of Baal, and the prophets of all the lesser gods, down to that river bank.' They took them down, and he slew them all. Then he went back to King Ahab. 'Get up,' he said. 'Eat and drink. Be quick: I hear the sound of rain.'

So Ahab went to eat and drink. And the rain came: the people of Israel had turned back to Yahweh, and the drought and famine were over.

Muhammad was born in the City of Mecca, in Arabia, in the year 570 of the Common Era. He became convinced that he had been called by Allah, the One True God, to be the Prophet of his people. The words which came to Muhammad from Allah are sacred to Muslims, and became the Qur'an, 'The Reading'. Is this account of Muhammad's Night Ride to Heaven a legend or an historical account? To many Muslims, it is literally true. Other Muslims would interpret it symbolically, and most non-Muslims would regard it as a legend.

The Night Journey

Muhammad was asleep in the bedroom of his home in Mecca when the archangel Jibra'il appeared.

'Arise, Muhammad,' the archangel said. 'The time is come.'

Like a man in a dream, Muhammad allowed himself to be led out into the garden, where he saw a white horse waiting. This was the fabulous Buraq, a horse so fast that it made time stand still.

The Prophet sat on Buraq's back, and in an instant he was transported from Mecca to Jerusalem, where he found himself in the mosque of El-Aqsa. Assembled in the mosque were all the prophets who had gone before, and Muhammad led them in a prayer of praise to Allah. He was then taken to visit the sacred rock, on which centuries before the prophet Abraham had made preparations to sacrifice his son Isaac to the Lord God.

From the rock, Muhammad was taken by Jibra'il to the seven Heavens of creation. At each Heaven, he met earlier prophets of other nations. He met Moses, prophet of the Children of Israel; he met Jesus, whom Muslims call Prophet of the Christians; and he met Abraham himself. In the Seventh Heaven, Muhammad met Adam, whom Muslims revere as the first prophet.

Then, past the Seventh Heaven, it seemed to Muhammad as if he had passed through a veil, and was able to see more than mortal eyes can see, more than mortal minds can imagine. Time stood still for him; he was in the presence of Allah.

Creation moved on. To his surprise, Muhammad found that he was back in Mecca. He sat quietly until dawn remembering Allah; then at sunrise, he made his way as usual to the Hanam Mosque.

Abu Jahl, an old rival of Muhammad, came up to him and smiled sarcastically.

'What have you to tell us today, Muhammad?' he enquired. 'What new thing has happened to you now?'

Muhammad described his night ride to Jerusalem and his ascension to the heavens. He said nothing of what had happened beyond the Seventh Heaven, because he could find no words to describe so sublime an experience.

Abu Jahl, shaken by Muhammad's quiet calm, sneered. 'How are we to believe all this? We've no one's word for it but yours.'

'I can prove it,' replied Muhammad. 'You know that until last night, I had never visited Jerusalem. Now, ask me anything you like about that city, and I will answer.'

And he proceeded to describe Jerusalem in such detail that even Abu Jahl was impressed. He tried another tack.

'What about the journey from Mecca to Jerusalem, on this amazing horse?' he enquired. 'Did you see anything on the way, or were you going too fast?'

'I will tell you exactly what I saw,' replied Muhammad. 'In the valleys outside Mecca I met a camel train camped for the night, and noticed that one of their camels had wandered away. I told them about this. Then I came upon another camel caravan, where I stopped for a drink of water from a jar which stood outside the largest of the tents. This camel train was heading for Mecca, and should reach the city sometime today. It is led by a dark grey camel carrying a double load, half wrapped in black cloth and half in a cloth of many colours.'

Led by Abu Jahl, the doubters set off to find the camel caravan Muhammad had described so exactly; soon they came across one which fitted the Prophet's description in every detail. They asked the camel drivers about the jar of water, and this was confirmed; they even found the second camel train, and were told that a camel had indeed escaped that night, and that a strange voice had been heard warning them of this.

The unbelievers in Mecca had failed to cast doubt into the minds of Muhammad's followers. They were taken aback by the Prophet's knowledge of the El Aqsa mosque in Jerusalem and the proofs of his meeting with the camel trains on the way to Jerusalem. Because they could not disprove Muhammad's story, their hostility grew towards Muhammad and his followers. Soon, the Prophet and his followers, called 'Muslims', had to leave Mecca, and find new homes across the world.

In the sixteenth century of the Common Era, northern India was overrun by the Moghuls. They converted millions to their faith, Islam, at the point of the sword. The leaders of the Sikh religion, the Gurus, always stood up for the freedom of the individual, and many of them were martyred (that is, killed for their beliefs). One of the Gurus, Guru Gobind Rai, decided that the time had come for Sikhs to stand up and fight for their beliefs. The story that follows dates from 1699. No-one can say precisely what took place, but many Sikhs believe that it happened exactly as recounted here.

Guru Gobind Rai and the Formation of the Khalsa

Gobind Rai thought for a long time about the crisis that was facing his followers. He was opposed to the political tyranny of the Moghuls, but he was also opposed to the old social and religious tyranny of the rigid Hindu caste system, in which the priestly caste lorded it over all others. On the occasion of the Baisakhi celebrations, the April Harvest Festival of the year 1699, Gobind Rai called together Sikhs from all over India to the Punjab. Eighty thousand Sikhs responded to the Guru's call.

A large tent was pitched in front of the assembly, and when everyone had arrived and had sat down, Guru Gobind Rai emerged from the tent and stood on a dais in front of the gathering with a drawn sword.

'My faithful Sikhs,' the Guru cried. 'Is there anyone here who would lay down his life for his beliefs? I want the head of a Sikh. I must have a sacrifice!'

And he brandished the sword above his head fiercely.

There was a murmur of astonishment among the assembled ranks, but no response. The Guru repeated his challenge, and again no-one moved.

At the third call, a man called Daya Ram stepped forward and bowed to the Guru.

'My Lord,' he said. 'My head is yours, to dispose of as you will.'

Gobind Rai led him into the tent. The assembled Sikhs rose

to their feet expectantly. There was a swish, a thud, and those nearest the tent saw a stream of blood coming from under the canvas.

The Guru emerged with his great sword dripping fresh blood. He held it aloft, and the people shrank back.

'I want another head!' shouted Gobind. 'My sword wants to taste the blood of another beloved Sikh. We must sacrifice to gain our freedom!'

At this, there was much muttering and stirring among the people, and many began to slink away. But one stood up and came forward – Dharam Das, of Delhi. He too offered his head to the Guru, and was led into the tent.

Again came the swish and thud, and once more blood flowed.

The Guru came out again with the same demand, and one after another three more Sikhs came forward – Mokham Chand, Sahib Chand and Himmat Rai. One by one they faced the ordeal of the dripping sword. Five brave Sikhs had faced the Guru's supreme test.

Then a remarkable thing happened. Before the eyes of the astonished assembly, the Guru came out of the tent – accompanied by the same five Sikhs! They were wearing new uniforms, and glowing with new confidence.

The Guru, helped by his five 'beloved ones', prepared an iron pot of 'amrit', a nectar made from sugar cakes and water, stirred with a khanda or double-edged sword. Gobind Rai then gave some of the amrit to each of his five disciples; in this way he declared them all equal, and so abolished the hated caste system.

The five were given a new surname, Singh, added to their names. The Guru himself received the amrit from the five, and took the name Guru Gobind Singh. Singh means 'brave as a lion'. Soon the Guru had an army of followers, all prepared to die for their faith; all had taken the amrit and assumed the surname Singh. This army became known as the order of the Khalsa – the army of soldier saints.

BIBLIOGRAPHY

POLYNESIA AND AUSTRALIA
Grimble, Sir Arthur (1952) *A Pattern of Islands*, John Murray
Poignant, Roselyn (1968) *Oceanic Mythology*, Hamlyn

NORTH AMERICAN INDIAN
Burland, C. (1968) *North American Indian Mythology*, Hamlyn

AFRICA
Arnott, Kathleen (1962) *African Myths and Legends*, OUP
Carey, Margaret (1970) *Myths and Legends of Africa*, Hamlyn
Parrinder, Geoffrey (1967) *African Mythology*, Hamlyn
Parrinder, Geoffrey (1976) *African Traditional Religion*,
   Harper & Row/Sheldon Press

EGYPT
Divin, Marguerite (1965) *Stories from Ancient Egypt*, Burke
Green, Roger Lancelyn (1972) *Tales of Ancient Egypt*, Puffin
Ions, Veronica (1969) *Egyptian Mythology*, Hamlyn

SUMERIA
Grey, J. (1969) *Near Eastern Mythology*, Hamlyn
Sandars, N. K. (1970) *The Epic of Gilgamesh*, Penguin

IRAN
Hinnells, J. R. (1974) *Persian Mythology*, Hamlyn

ISRAEL
The New English Bible (1972) OUP/CUP
The Bible: Revised Standard Version (1973) Bible Society
Good News Bible: Today's English Version (1976) Bible
   Society
Dale, Alan (1973) *The Winding Quest*, OUP

ISLAM
Boyce, R. (1972) *The Story of Islam*, REP
Pickthall, Mohammed M. (1930) *The Meaning of the Glorious
   Koran*, Allen and Unwin/Mentor

GREECE
Grant, Michael (1964) *Myths of the Greeks and Romans*, Mentor
Graves, Robert (1955) *The Greek Myths*, 2 vols., Pelican
Pinsent, J. (1969) *Greek Mythology*, Hamlyn
Rose, H. J. (1959) *A Handbook of Greek Mythology*, Dutton

SCANDINAVIA AND NORTHERN EUROPE

Davidson, H. R. Ellis (1965) *Gods and Myths of Northern Europe*, Penguin

Davidson, H. R. Ellis (1969) *Scandinavian Mythology*, Hamlyn

Nye, R. (1972) *Beowulf: Bee Hunter*, Faber

Picard, Barbara Leonie (1953) *Tales of the Norse Gods and Heroes*, OUP

Sutcliffe, Rosemary (1970) *Dragon Slayer*, Bodley Head

EUROPE — CHRISTIANITY

Every, G. (1970) *Christian Mythology*, Hamlyn

Green, Roger Lancelyn (1970) *King Arthur & His Knights of the Round Table*, Puffin

Malory, Thomas trs. Matarasso, P. M. (1969) *The Quest of the Holy Grail*, Penguin

INDIA — HINDUISM

Dutt, Romesh C. (1969) *The Ramayana and the Mahabharata*, Dent

Gray, J. E. (1979) *Indian Tales and Legends*, OUP

Ions, Veronica (1967) *Indian Mythology*, Hamlyn

Ions, Veronica (1970) *Myths and Legends of India*, Hamlyn

O'Flaherty, Wendy (1975) *Hindu Myths*, Penguin

INDIA — SIKHISM

McLeod, W. H. (1975) *Way of the Sikh*, Hulton

McLeod, W. H. (1971) *The Sikhs of the Punjab*, Oriel Press

CHINA

Birch, Cyril (1961) *Chinese Myths and Fantasies*, OUP

Christie, A. (1968) *Chinese Mythology*, Hamlyn

Van Owen, Raymond (1973) *Taoist Tales*, Mentor

GENERAL

Barber, Richard (1979) *A Companion to World Mythology*, Kestrel

Farmer, Penelope (1978) *Beginnings: Creation Myths of the World*, Chatto

Parrinder, E. G. (1971) *Man and his Gods: an Encyclopaedia of the World's Religions*, Puffin

Robinson, H. S. and Wilson, K. (1967) *Encylopaedia of Myths and Legends of all Nations*, Kaye & Ward.